Axioms for Lattices and Boolean Algebras

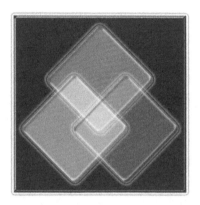

Axioms for Lattices and Boolean Algebras

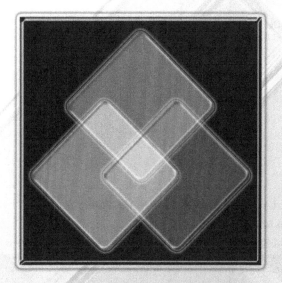

R Padmanabhan
University of Manitoba, Canada

S Rudeanu
University of Bucharest, Romania

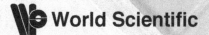 World Scientific

NEW JERSEY · LONDON · SINGAPORE · BEIJING · SHANGHAI · HONG KONG · TAIPEI · CHENNAI

Published by

World Scientific Publishing Co. Pte. Ltd.

5 Toh Tuck Link, Singapore 596224

USA office: 27 Warren Street, Suite 401-402, Hackensack, NJ 07601

UK office: 57 Shelton Street, Covent Garden, London WC2H 9HE

British Library Cataloguing-in-Publication Data
A catalogue record for this book is available from the British Library.

AXIOMS FOR LATTICES AND BOOLEAN ALGEBRAS

ISBN-13 978-981-283-454-6
ISBN-10 981-283-454-0

Printed in Singapore.

Introduction

Ever since Euclid the axiomatic approach is at the heart of mathematics. The axiomatic approach admits the possibility of a mixture of deductive and empirical reasoning and hence it is an ideal pedagogical tool. Also, in the emerging 21st century it is the natural choice of modern theorem-provers for development and experimentation of automated reasoning. Among the various types of axioms one can formulate for a given theory, identities are the most natural ones. Many familiar algebraic systems occurring in lattice theory are usually defined by means of equational identities, i.e., sentences in the form $f = g$, where f and g are formed from variables and symbols denoting the fundamental operations of the relevant algebras. The purpose of this monograph is to collect and present all known minimal equational bases for semilattices, lattices, modular lattices and Boolean algebras.

There is a huge literature on the axioms of several equational classes of lattices from 1880 onwards. The original 1963 monograph by Rudeanu – the genesis of this monograph – reports the state of the art of the subject at that time. The present book updates the original monograph in several respects. We report not only new axiom systems – and there are a lot! – but also several deep metatheorems (i.e., theorems about axiom systems) that have been proved in the meantime. Unlike the old monograph, the present one includes many proofs. Besides, the strategy in presenting old papers has been changed. Let us explain all this in some detail.

The first four chapters of the book present systems of axioms for semilattices and lattices, modular lattices, distributive lattices, and Boolean algebras and orthomodular lattices, no matter whether they are given in terms of join and meet or in terms of other tools, such as a ternary operation, a ternary relation, or others. In the case of Boolean algebras most systems use complementation (either as a basic operation or in a disguised form) together with join and meet or with join only, while other systems define Boolean algebras in terms of a single binary operation (the "Sheffer stroke") or in terms of ring operations; we have included all of them, as well as the axioms of the related concept of a Boolean group. A few related lines of research are sketched in Chapter 5, where we also suggest several open problems.

In order to keep the dimension of our book within reasonable limits, we have in general not included systems that characterize a class of lattices within a larger class. In other words, we have almost exclusively collected systems that do not reproduce all the axioms of a class of lattices larger than the one being defined. The most notable exceptions are a few characterizations of modular lattices (Chapter 2, §1) and of distributive lattices (Chapter 3, §1), which are immediately obtained from the definitions, and a major exception, the characterization of Boolean algebras within the class of all uniquely complemented lattices (Chapter 4, §8). Here we address the celebrated problem of E.V. Huntingron (1904), which, according to a leading expert in modern lattice theory, is one of the two problems that shaped a century of research in lattice theory (cf. George Grätzer). The problem was whether every uniquely complemented lattice is distributive. Huntington believed that every such lattice was distributive, and hence Boolean. He himself gave some sufficient conditions which force a uniquely complemented lattice to be Boolean. Birkhoff and von Neumann proved that modularity is one such property, i.e., every uniquely complemented modular lattice is distributive. Although Huntington's conjecture was disproved by Dilworth in 1945 (he established that every lattice is isomorphic with a sublattice of a lattice with unique complements), the interest for finding conditions which ensure distributivity of uniquely complemented lattices has remained intact. In Chapter 4 we show that there are uncountably many non-modular lattice identities which force a uniquely complemented lattice to be Boolean, thus providing several new axiom systems for Boolean algebras within the class of all uniquely complemented lattices.

We have included, to the best of our knowledge, all of the papers that suit the above description. We have actually described in the book the most significant systems in our appreciation; this has resulted, for instance, in more than 40 systems for distributive lattices or bounded distributive lattices and 70 systems for Boolean algebras. However, certain authors provided hundreds or thousands of axioms systems (!), as shown in the book. Some of the systems we have chosen are given with proofs and some of the proofs in the book are new, without mention of this fact.

Beside the five chapters, our monograph comprises six appendices. Appendix A, written by W. McCune, reproduces four proofs provided by a computer program called Prover9. Appendix B is a bibliography on axiom systems for partially ordered systems and betweenness in posets. Appendix C is a very short presentation of quasilattices, a class of algebras $(A, \vee, \wedge)$ that captures the essence of all regular identities in lattices. Appendix D compiles a bibliography of papers devoted to the axiomatics of Lukasiewicz-Moisil algebras, which play an important role in algebraic logic. Appendix E lists a few papers which suggest methods for testing the associativity of a binary operation. Several papers that deal with E.H. Moore's complete existential theory (i.e., a kind of exhaustive analysis of all existing implications that link a set of conditions) are briefly presented in Appendix F.

The Bibliography refers to the four chapters, while each appendix has its own bibliography.

While the 1963 monograph also aimed at exhaustiveness in the sense described above, the present monograph has several new features. One of them was already mentioned: the inclusion of many proofs.

A manifest tendency in the literature is the search for minimal equational bases (i.e., systems of equational identities with the smallest possible number of axioms) for equational classes of lattices and in particular the search for definitions by a single equational identity whenever such single identities exist. A new feature in this monograph is the application of some structural properties in discovering minimal equational bases. Thus, for example, while the definability of groups by a single axiom depends upon the type in which groups are defined (cf. Tarski-Green, 1968), all finitely-based varieties of orthomodular lattices (in particular, Boolean algebras) are always one-based, whatever the type, thanks to the permutable and distributive congruence properties.

Another new feature is to exploit the self-dual nature of the classes of lattices dealt with in order to give independent self-dual sets of axioms.

Last but not least, another new feature of this monograph is the emphasis on the computer program Otter and its improved version Prover9, which turn out to be very efficient theorem-provers.

At this point certain technical explanations seem necessary.

The papers referred to in this book cover a period of more than hundred years. As the mathematical style has meanwhile altered to a certain extent, the realization of a unitary presentation in our book raised certain problems. Thus in very old papers axioms were not understood just as properties of the operations, but they also included the very existence of these operations;

for instance, any system of axioms for lattices began with something like "With any $x, y \in A$ is associated an element $x \vee y \in A$" and a similar axiom for $\wedge$. Nowadays the role of such "existence axioms", as they were called, is played by the genus proximus, which is an algebra of a certain signature. In our book we refer to these early papers as if they had been written according to contemporary standards, like "A lattice is an algebra $(A, \vee, \wedge)$ of type $(2,2)$ such that ..." followed by the remaining axioms of the author. This strategy is exactly the opposite of the one adopted in the monograph by Rudeanu [1963].

A curiosity of certain old papers is that they include pairs of axioms of the form p and $p \implies q$, instead of simply listing axioms p and q. This was probably due to the desire of facilitating the proof of the independence of axiom p : any model in which axiom p fails automatically satisfies axiom $p \implies q$. We have taken the liberty of ignoring these artifices, by replacing axiom $p \implies q$ by axiom q. To be sure, in all these circumstances we say nothing about the independence of the system, even if the original system was independent.

For each type of statement, the displayed statements of this book are numbered separately according to the usual Statement $m.n.p$ convention. For instance, Proposition $m.n.p$ means the p–th proposition in Section n of Chapter m. However within Chapter m the statements $m.n.p$ may be referred to simply as $n.p$.

Formulae are numbered in each section by a single number; formula $(n.p)$ means formula (p) in Section n of the current chapter, while $(m.n.p)$ is formula $(n.p)$ in Chapter m.

The notation Author [year]* indicates papers that are not available in our libraries and which we quote from other sources, mainly from Mathematical Reviews.

Acknowledgements. We sincerely thank George Grätzer for the keen interest he took in this project ever since it was conceived. Needless to say that R. Padmanabhan benefited a lot by periodically presenting his discoveries in Dr. Grätzer's seminar over the past 40 years. We also thank George for inserting a pre-publication announcement of this book in his forthcoming survey on lattice theory "Two problems that shaped a century of lattice theory". We thank Dr. David Kelly for reading the Introduction and making some constructive comments which enhanced our presentation. R. Padmanabhan also thanks all his collaborators (David Kelly, Harry Lakser, William McCune, Craig Platt, R.W. Quackenbush,

Robert Veroff and Barry Wolk) for having stimulating discussions on lattice axioms over the years and to Dr. Guenter Krause, Head of the Department of Mathematics, University of Manitoba for creating a pleasant atmosphere and camaraderie conducive for doing productive research. We thank Bill McCune for sending us a tex-formatted Prover9 proofs which form Appendix A. S. Rudeanu wishes to thank Dan A. Simovici, who during several years urged him to write this book. We also thank Laurenţiu Leuştean, who provided us with quite a lot of papers.

The authors deeply appreciate the editorial staff at World Scientific for their meticulous attention to details and we express our sincere gratitude to Ms. Lai Fun Kwong for her expert advice and for the careful proof-reading.

R. Padmanabhan's research was supported by an ongoing operation grant from NSERC of Canada for the past 38 years.

Last but certainly not least, we are extremely grateful to Cristian S. Calude for his kind appreciation of our book, for his efforts in promoting it and for his invaluable "TeXnical" help, without which, our monograph would have never made it this far. Saying "Thank you, Cris" would be a massive understatement.

Contents

Introduction   v

1. Semilattices and Lattices   1

2. Modular Lattices   39

3. Distributive Lattices   53

4. Boolean Algebras   69

5. Further Topics and Open Problems   137

Appendix A: Some Prover9 Proofs   147

Appendix B: Partially Ordered Sets and Betweenness   159

Appendix C: Quasilattices   181

Appendix D: Lukasiewicz-Moisil Algebras   183

Appendix E: Testing Associativity   187

Appendix F: Complete Existential Theory and Related Concepts   189

Bibliography   193

Index   211

Chapter 1

Semilattices and Lattices

A *lattice* is a structure $(L, \leq)$, where $\leq$ is a partial order on L, that is, it satisfies the properties

P_1 $\qquad \forall x \, (x \leq x)$,

P_2 $\qquad \forall x \, \forall y \, (x \leq y \, \& \, y \leq x \Longrightarrow x = y)$,

P_3 $\qquad \forall x \, \forall y \, \forall z \, (x \leq y \, \& \, y \leq z \Longrightarrow x \leq z)$,

(*reflexivity, antisymmetry* and *transitivity*), and every two elements have a *least upper bound* and a *greatest lower bound*:

P_4 $\qquad \forall x \, \forall y \, \exists s \, (x \leq s \, \& \, y \leq s \, \& \, (x \leq z \, \& \, y \leq z \Longrightarrow s \leq z))$,

P_5 $\qquad \forall x \, \forall y \, \exists p \, (p \leq x \, \& \, p \leq y \, \& \, (z \leq x \, \& \, z \leq y \Longrightarrow z \leq p))$.

It is easily shown that the elements s and p are uniquely determined by x and y, which enables one to define

(D) $$x \vee y = s \, , \; x \wedge y = p \, .$$

The system of axioms $\{P_1, P_2, P_3, P_4, P_5\}$ is due to Peirce [1880-84] and Ore [1935], who corrected certain flaws in the original system of Peirce. Sorkin [1951] proved the independence of the system. Another system is due to Bennett [1930], who used axioms P_1, P_2, P_3 and two more complicated variants of P_4 and P_5.

More generally, a structure $(L, \leq)$ is called a *join semilattice* if it satisfies P_1, P_2, P_3, P_4, and is known as a *meet semilattice* if it satisfies P_1, P_2, P_3, P_5.

Since the elements s and p in (D) are uniquely determined by x and y, this opens the way to the definition of semilattices and lattices as algebras, whose axioms will be studied in §1 and §§2,3, respectively. The concept of identity, which is crucial in this book, is carefully explained in §1 in the particular case of groupoids. In §4 we present systems of axioms for lattices in terms of lattice betweenness, segments or partially defined ternary operations.

1

1.1. Semilattices

It is easily seen that the operation $\vee$ of a join semilattice and the operation $\wedge$ of a meet semilattice are *idempotent, commutative* and *associative*. This remark has led to the concept of semilattice as a groupoid. Recall that a *groupoid* or an *algebra of type* (2) $(S, \circ)$ is a set S endowed with a binary operation $\circ : S \times S \longrightarrow S$. So a *semilattice* is a groupoid $(S, \circ)$ which satisfies the identities

S_1 $\quad x \circ x = x$,

S_2 $\quad x \circ y = y \circ x$,

S_3 $\quad (x \circ y) \circ z = x \circ (y \circ z)$.

This concept is due to Klein-Barmen [1934]. See also the historical notice in Birkhoff [1948], Ch.II, footnote 6, but note that the name Halbverband had been used earlier, in Klein-Barmen [1939].

The independence of the system of axioms $\{S_1, S_2, S_3\}$ was established e.g. in Sorkin [1951], Dubreil-Jacotin, Lesieur and Croisot [1953] and Rudeanu [1959].

It is immediately seen that the following identities hold in a semilattice:

S_4 $\quad (u \circ v) \circ ((w \circ x) \circ (y \circ z)) = ((u \circ v) \circ (x \circ w)) \circ (z \circ y)$,

S_5 $\quad ((x \circ y) \circ z) \circ t = t \circ (x \circ (y \circ z))$,

S_6 $\quad (x \circ y) \circ z = (y \circ z) \circ x$,

S_7 $\quad x \circ (y \circ z) = z \circ (x \circ y)$.

They have been used in several two-axiom characterizations of semilattices, namely $\{S_1, S_4\}$, $\{S_1, S_5\}$, $\{S_1, S_6\}$ and $\{S_1, S_7\}$, due to Potts [1965], Petcu [1967], Padmanabhan [1966] and Sobociński [1979], respectively.

Let us prove that the last two systems actually define semilattices. Since S_6 and S_7 reduce to S_3 in presence of commutativity, it suffices to prove S_2. It follows from S_1 and S_6 that

$$y \circ z = (y \circ z) \circ (y \circ z) = (z \circ (y \circ z)) \circ y = ((y \circ z) \circ y) \circ z$$

$$= ((z \circ y) \circ y) \circ z = ((y \circ y) \circ z) \circ z = (y \circ z) \circ z = (z \circ z) \circ y = z \circ y ,$$

while S_1 and S_7 imply

$$y \circ z = y \circ (z \circ z) = z \circ (y \circ z) = z \circ ((y \circ y) \circ z) = z \circ (z \circ (y \circ y))$$
$$= z \circ (y \circ (z \circ y)) = (z \circ y) \circ (z \circ y) = z \circ y \,.$$

$\square$

Ruedin [1966], [1966/67] remarked that the system $\{S_1, S_2, S_3\}$ can be replaced by $\{S_1, S_2, S_8, S_9\}$, where

S_8 $x \circ (y \circ x) = y \circ x \,,$

S_9 $x \circ (y \circ z) = (x \circ y) \circ (x \circ z) \,;$

the proof is a refinement of the well-known proof that transforms $(S, \circ)$ into a join/meet semilattice. From this Ruedin derived a characterization of semilattices with neutral element, previously obtained by Felscher and Klein-Barmen [1959].

There is one more equivalent definition of semilattices. To state it we need a few preliminaries.

Let $(S, \circ)$ be a groupoid and let V be a set of elements called *variables*. The set $V \cup \{\circ, (,)\}$ is called *alphabet* and its elements are said to be *letters*; a *word* is a concatenation of letters. The set of $\circ-$*expressions* or $\circ-$*terms* is the least set of words obeying the following rules: 1) every variable is a $\circ-$term, and 2) if φ and ψ are $\circ-$terms, then $(\varphi) \circ (\psi)$ is a $\circ-$term; however we write x instead of (x) for every variable x occurring in a compound expression. In universal algebraic terminology, the $\circ-$terms are simply the elements of the *clone* $< \circ >$ generated by $\circ$. Thus every term φ is obtained in finitely many steps by applying the above rules 1) and 2). The variables $x_1, \ldots, x_n$ occurring in this construction of φ are known as the *variables of* φ; we also say that φ is an expression *in the variables* $x_1, \ldots, x_n$.

The crucial point is that each $\circ-$term φ *generates a function* with arguments and values in S, which is obtained by interpreting each letter $x \in V$ occurring in φ as a variable in the usual sense, and each occurrence of the letter $\circ$ as the symbol of the binary operation of the algebra $(S, \circ)$. The functions generated by terms are called *term functions* or *polynomials*.

An identity is currently meant as something like

$$\forall x_1 \ldots \forall x_n \; f(x_1, \ldots, _n) = g(x_1, \ldots, x_n) \,;$$

however in axiomatics we refer to identities that may or may not be fulfilled. The exact meaning of this alternative is the following. Whenever the concept of expression is naturally defined over an algebra A like in the above particular case of groupoids, by an *identity* $\varphi = \psi$ is meant in fact a notation for a pair (φ, ψ) of terms φ, ψ; we say that A *satisfies* this identity if the terms φ, ψ generate the same polynomial.

Now we are going to prove the promised result.

Lemma 1.1.1. *In a semilattice $(S, \circ)$ every $\circ$-term in n variables $x_1, \ldots, x_n$ generates the function $x_1 \circ \cdots \circ x_n$.*

PROOF: By algebraic induction on the expression φ. If φ is a variable x_1, the property is trivial. Suppose $\varphi = \alpha \circ \beta$, where the terms α and β satisfy the property. Let $y_1, \ldots, y_p$ and $z_1, \ldots, z_q$ be the variables of α and β, respectively. Then $\{y_1, \ldots, y_p\} \cup \{z_1, \ldots, z_q\} = \{x_1, \ldots, x_n\}$ and $\alpha \circ \beta = (y_1 \circ \cdots \circ y_p) \circ (z_1 \circ \cdots \circ z_q) = x_1 \circ \ldots x_n$ by S_1, S_2 and S_3.   □

Lemma 1.1.2. *If the groupoid $(S, \circ)$ satisfies S_1, the polynomial f generated by a $\circ$-term satisfies $f(x, \ldots, x) = x$.*

PROOF: Again by algebraic induction. The inductive step follows from

$$(\alpha \circ \beta)(x, \ldots, x) = \alpha(x, \ldots, x) \circ \beta(x, \ldots, x) = x \circ x = x \,.$$

□

Theorem 1.1.1. (Petcu [1971]). *A groupoid $(S, \circ)$ is a semilattice if and only if every two $\circ$-expressions in the same variables generate the same function.*

We are going to prove a slight refinement of this theorem, which needs the following introduction. A polynomial in n variables is said to be *essentially n-ary* if it depends actually upon all of its n variables. Let $P_n(S)$ denote the number of essentially n-ary polynomials of a groupoid S.

Theorem 1.1.1' *The following conditions are equivalent for a groupoid $(S, \circ)$:*

(i) $P_n(S) = 1$ for all n;

(ii) $P_n(S) = 1$ for $n := 1, 2, 3$;

(iii) $(S, \circ)$ is a semilattice.

PROOF: (i)$\Longrightarrow$(ii): Trivial.

(ii)$\Longrightarrow$(iii): Applying $P_1(S) = 1$ we get $x \circ x = x$ because both x and $x \circ x$ are unary. Similarly, by applying $P_2(S) = 1$ we obtain commutativity since $x \circ y$ and $y \circ x$ must coincide, and finally $P_3(S) = 1$ forces associativity.

(iii)$\Longrightarrow$(i): By Lemma 1.1.   □

Recently G. Grätzer asked whether semilattices can be defined by systems containing identities of arbirary length. The affirmative answer is provided by the following theorem, which was proved by Padmanabhan and Wolk in the early 1970's, but has so far remained unpublished.

Theorem 1.1.2. *For any $n > 2$, ,S_1 and*

(SLn)
$$x_1 \circ (x_2 \circ (x_3 \circ \cdots \circ (x_{n-2} \circ (x_{n-1} \circ x_n))) \ldots)$$
$$= x_2 \circ (x_3 \circ \cdots \circ (x_{n-1} \circ (x_n \circ x_1)) \ldots) ,$$

form an independent system of axioms defining semilattices.

PROOF: The system $\{S_1, SL3\}$ defines semilattices because by applying SL3 twice we obtain $S_7 : x \circ (y \circ z) = y \circ (z \circ x) = z \circ (x \circ y)$. The rest of the proof consists in showing that $SLn \implies SL(n-1)$. To simplify notation we will use concatenation instead of $\circ$ and the shortcut $x_3(x_4(\ldots(x_{n-3}(x_{n-2} = Y$. So SLn reads

(1) $$x_1 (x_2 (Y (x_{n-1}x_n)) \ldots) = x_2 (Y (x_{n-1}(x_n x_1)) \ldots) .$$

By taking $x_n := x_{n-1} := x_1$ and using idempotency S_1 we obtain

(2) $$x_1 (x_2 (Y x_1) \ldots) = x_2 (Y x_1) \ldots) ,$$

and by taking further $x_{n-2} := x_{n-3} := \cdots := x_3 := x_1$ we get

(3) $$x_1 (x_2 x_1) = x_2 x_1 .$$

On the other hand, taking $x_1 := x_{n-1}x_n$ in (1) yields

(4) $$(x_{n-1}x_n)(x_2 (Y (x_{n-1}x_n)) \ldots) = x_2 (Y (x_{n-1} (x_n(x_{n-1}x_n))) \ldots) .$$

Now take $x_1 := x_{n-1}x_n$ in (2) and use in turn (4), (3), (1) and S_1; then

$$x_2 (Y (x_{n-1}x_n) \ldots) = (x_{n-1}x_n)(x_2 (Y (x_{n-1}x_n)) \ldots)$$

$$= x_2 (Y (x_{n-1} (x_n (x_{n-1}x_n)) \ldots) = x_2 (Y (x_{n-1} (x_{n-1}x_n)) \ldots)$$

$$= x_n (x_2 (Y (x_{n-1}x_{n-1})) \ldots) = x_n (x_2 (Y x_{n-1}) \ldots) ,$$

whence we obtain $SL(n-1)$ for $x_n := x_1$.

To prove the independence of SLn define $x \circ y = x$, while with $x \circ y = $ constant one obtains the independence of S_1. □

A similar question was answered by Tarski [1968], who proved that semilattices can be defined by independent systems of as many axioms as desired. His proof is purely existential and is based on topological methods. We give below a constructive proof.

First we introduce a notation. Given a groupoid $(S, \circ)$ we define

(5) $$< 1 > x = x, \quad < n+1 > x = x \circ < n > x, \quad n \in \mathbb{N} \backslash \{0\} .$$

Lemma 1.1.3. *Suppose $a, b \in \mathbb{N}\backslash\{0\}$ and identities $< a + 1 > x = < b + 1 > x = x$ hold. If $c = $ g.c.d.$\{a, b\}$ then $< c + 1 > x = x$.*

PROOF: Recall that the Euclidean algorithm for finding c can be given the following "subtractive" form[†]: Suppose e.g. that $a > b$. Define $c_0 = a$ and while $c_i \geq b$ set $c_{i+1} = c_i - b$. Then $c = c_q$, where q is the index such that $c_q < b$.

It follows that $< c_0 + 1 > x = x$ and if $i \leq q$ satisfies $< c_{i-1} + 1 > x = x$ then

$$< c_i + 1 > x = x \circ < c_i > x = < b + 1 > x \circ < c_{i-1} - b > x$$

$$= < b + 1 + c_{i-1} - b > x = < c_{i-1} + 1 > x = x \,.$$

Therefore $< c_i + 1 > x = x$ $(i = 0, 1, \ldots, q)$, implying $< c + 1 > x = x$. □

Theorem 1.1.3. *For every $n \geq 2$ there is an independent system of n identities defining semilattices.*

PROOF: For $n := 2$ we have already given several systems of two axioms, whose independence is easy to establish.

Now for every $n \geq 2$ we are going to construct an independent system of $n + 1$ identities, the last one being S_7.

Let $p_1, p_2, \ldots, p_n$ be the first n primes and $P = p_1 \ldots p_n$. Define $q_i = P/p_i$ $(i = 1, \ldots, n)$. We claim that

(6) $\{< q_i + 1 > x = x \ (i = 1, \ldots, n), \ x \circ (y \circ z) = z \circ (x \circ y)\}$

is an independent system defining semilattices.

Set $r_1 = q_1$ and $r_i = $ g.c.d.$\{r_{i-1}, q_i\}$ $(i = 2, \ldots, n)$. Then $< r_1 + 1 > x = x$ and if $< r_{i-1} + 1 > x = x$ then Lemma 1.3 implies $< r_i + 1 > x = x$. Therefore $< r_i + 1 > x = x$ for $i = 1, \ldots, n$; in particular $< r_n + 1 > x = x$. But according to a well-known property, $r_n = $ g.c.d.$\{q_1, \ldots, q_n\} = 1$, therefore $< 2 > x = x$, which is S_1.

To prove independence, take first $S := \mathbb{Z}_3 = \{0, 1, 2\}$ and $x \circ y = 2x + 2y$, where the operations are taken modulo 3. Then the commutative operation $\circ$ is not associative, because $(x \circ y) \circ z = x + y + 2z$, while $x \circ (y \circ z) = 2x + y + z$. But $x \circ x = x$, hence $< m > x = x$ for all m. Therefore axiom S_7 is independent.

[†]Useful in computer science!

To proceed further, for each $i \in \{1, \ldots, n\}$ consider the field $(\mathbb{Z}_{p_i}, +, \cdot)$, where $\mathbb{Z}_{p_i} = \{\overline{0}, \ldots, \overline{p_i - 1}\}$ and the operations are taken modulo p_i. Then $<m> x = x + \cdots + x = \overline{m} \cdot x$, therefore

$$<m> x = \overline{0} \Longleftrightarrow \overline{m} = \overline{0} \Longleftrightarrow p_i \mid m .$$

Consequently $<q_i> x \neq \overline{0}$, hence $<q_i + 1> x \neq x$, whereas for every $j \neq i$ we have $<q_j + 1> x = <q_j> x + x = x$. Therefore each axiom $<q_i + 1> x = x$ is independent. $\square$

The above theorem also provides a new answer to the question addressed in the previous theorem: the lengths of identities in independent bases for semilattices may be as large as one pleases. So these are "unbounded" in every sense of the term.

1.2. Defining Lattices in Terms of the Operations $\vee$ and $\wedge$

It is easily seen that the operations $\vee$ (*join*) and $\wedge$ (*meet*) of a lattice (cf. P_4, P_5 and (1)) are *idempotent, commutative, associative* and satisfy the *absorption laws*, i.e.,

$L_1^\vee$	$x \vee x = x$,
$L_1^\wedge$	$x \wedge x = x$,
$L_2^\vee$	$x \vee y = y \vee x$,
$L_2^\wedge$	$x \wedge y = y \wedge x$,
$L_3^\vee$	$(x \vee y) \vee z = x \vee (y \vee z)$,
$L_3^\wedge$	$(x \wedge y) \wedge z = x \wedge (y \wedge z)$,
$L_4^\vee$	$x \vee (x \wedge y) = x$,
$L_4^\wedge$	$x \wedge (x \vee y) = x$.

Conversely, if an algebra $(L, \vee, \wedge)$ satisfies the above properties, then it can be proved that the equivalence

$$(1) \qquad\qquad x \vee y = y \Longleftrightarrow x \wedge y = x$$

holds and the relation $\leq$ defined by

$$(2) \qquad\qquad x \leq y \Longleftrightarrow x \vee y = y \,(\Longleftrightarrow x \wedge y = x)$$

satisfies properties $P_1 - P_5$. Thus a lattice is equivalently defined as an algebra $(L, \vee, \wedge)$ satisfying the system of axioms

$$\mathbf{L}_0 = \{L_1^\vee, L_1^\wedge, L_2^\vee, L_2^\wedge, L_3^\vee, L_3^\wedge, L_4^\vee, L_4^\wedge\} \ .$$

So, while in §1 we have regarded semilattices as algebras of type (2), in this section we deal with lattices defined as algebras *of type* (2,2), that is, endowed with two binary operations $\vee, \wedge$.

The separation of the system $\mathbf{L}_0$ from the other properties of a Boolean algebra was first accomplished by Schröder [1890-1905]. Then Dedekind [1897] noted that axioms $L_1^\vee, L_1^\wedge$ can be proved from $L_4^\vee, L_4^\wedge$:

$$x \vee x = x \vee (x \wedge (x \vee x)) = x \ , \ \ x \wedge x = x \wedge (x \vee (x \wedge x)) = x \ ,$$

so that system $\mathbf{L}_0$ is equivalent to

$$\mathbf{L}_1 = \{L_2^\vee, L_2^\wedge, L_3^\vee, L_3^\wedge, L_4^\vee, L_4^\wedge\} \ .$$

This was proved by Ore [1935]. The problem of the independence of $\mathbf{L}_1$ was raised by Birkhoff [1948] and solved in the affirmative by Kimura [1950].

The notation $L_i^\vee, L_i^\wedge$ emphasizes the pairs of *dual axioms*, i.e., which are obtained from each other by interchanging $\vee$ and $\wedge$. We denote by L_i the set $\{L_i^\vee, L_i^\wedge\}$. Thus, e.g., we have just proved that L_4 implies L_1.

A system of axioms is called *self-dual* if it consists of pairs of dual axioms; for instance, both $\mathbf{L}_0$ and $\mathbf{L}_1$ are self-dual. The existence of a self-dual system of axioms for a class of lattices implies the fact that the *principle of duality* holds for that class. This means that for each theorem valid in the lattices of that class, the *dual theorem* also holds: it is obtained from the original theorem by interchanging $\vee$ and $\wedge$.

Sorkin [1951] inaugurated a direction of research aimed at the idea of an exhaustive exploration of many alternatives. He considered all the variants of the absorption laws that can be obtained by permuting the letters x, y, namely $L_4^\vee, L_4^\wedge$ and

$L_5^\vee \qquad x \vee (y \wedge x) = x \ ,$

$L_5^\wedge \qquad x \wedge (y \vee x) = x \ ,$

$L_6^\vee \qquad (x \wedge y) \vee x = x \ ,$

$L_6^\wedge \qquad (x \vee y) \wedge x = x \ ,$

$L_7^\vee \qquad (y \wedge x) \vee x = x \ ,$

$L_7^\wedge \qquad (y \vee x) \wedge x = x \ ,$

and determined all the independent systems of axioms for lattice theory that can be obtained from the larger set of axioms

$$\Lambda_1 = \{L_2^\vee, L_2^\wedge, L_3^\vee, L_3^\wedge, \ldots, L_7^\vee, L_7^\wedge\} \; ;$$

in other words, all the independent subsystems of Λ_1 that are equivalent to Λ_1. There are 34 such systems, each of them having 6 or 7 axioms; one of these systems is of course the standard system $\mathbf{L}_1$.

Kalman [1951] generalized Sorkin's results as follows. For each of the 2^{14} subsystems λ of Λ_1 he found all the subsystems of Λ_1 that are equivalent to λ and all the axioms in Λ_1 that are implied by λ. This has applications to the theory of *skew lattices*. Roughly speaking, a skew lattice is an algebra $(L, \vee, \wedge)$ which has lattice-like properties, except the commutativity of the two operations.

According to a well-known theorem of Birkhoff, a class of algebras can be defined by a system of identities if and only if it is closed under the formation of subalgebras, direct products and homomophic images. Such classes of algebras are said to be *equational classes* or *varieties* ; for instance, the class of all lattices is equational. The above result explains the interest of finding equational characterizations of certain classes of algebras which have not originally been defined by identities, as was the case e.g. of modular lattices and Post algebras.

Note that all of the axioms $L_1, \ldots, L_7$ are identities. But the above comments do not exclude the interest of including axioms that are not identities. Thus Sorkin [1951] introduced the following weakenings of the absorption laws L_4 and L_7:

$L_8^\vee$ $x \wedge y = y \Longrightarrow x \vee y = x$,

$L_8^\wedge$ $x \vee y = y \Longrightarrow x \wedge y = x$,

$L_9^\vee$ $x \wedge y = x \Longrightarrow x \vee y = y$,

$L_9^\wedge$ $x \vee y = x \Longrightarrow x \wedge y = y$,

respectively, and determined all the independent subsystems of

$$\Lambda_2 = \{L_1^\vee, L_1^\wedge, L_2^\vee, L_2^\wedge, L_3^\vee, L_3^\wedge, L_8^\vee, L_8^\wedge, L_9^\vee, L_9^\wedge\}$$

that define lattices. There are 18 such systems, each of them having 7 axioms; one of them is a system introduced by Klein-Barmen [1932] and whose independence was first proved by Kobayasi [1943].

Rudeanu [1959] solved the same problem for the larger set

$$\Lambda_3 = \Lambda_2 \cup \{L_4^\vee, L_4^\wedge\}$$

and refound, of course, all the systems determined by Sorkin, plus 22 new systems, each of them having 6 or 7 axioms.

Petcu [1964] proved that lattices can be defined using only variants of the absorption and associative laws. Namely, he considered the axioms $L_3^\vee, L_3^\wedge$ and

$L_{10}^\vee$ $\qquad (x \vee y) \vee z = x \vee (z \vee y)$,

$L_{11}^\vee$ $\qquad (x \vee y) \vee z = y \vee (x \vee z)$,

$L_{12}^\vee$ $\qquad (x \vee y) \vee z = y \vee (z \vee x)$,

$L_{13}^\vee$ $\qquad (x \vee y) \vee z = (y \vee z) \vee x$,

$L_{14}^\vee$ $\qquad (x \vee y) \vee z = (z \vee y) \vee x$,

$L_{15}^\vee$ $\qquad (x \vee y) \vee z = (x \vee z) \vee y$,

$L_{16}^\vee$ $\qquad x \vee (y \vee z) = y \vee (x \vee z)$,

$L_{17}^\vee$ $\qquad x \vee (y \vee z) = z \vee (x \vee y)$,

$L_{18}^\vee$ $\qquad x \vee (y \vee z) = z \vee (y \vee x)$,

and their duals $L_{10}^\wedge, \ldots, L_{18}^\wedge$. Petcu showed that these are all the distinct variants of the associative law that can be imagined up to a permutation of the variables x, y, z, and looked for independent subsystems of

$$\Lambda_4 = \{L_4^\vee, L_4^\wedge, \ldots, L_7^\vee, L_7^\wedge, L_3^\vee, L_3^\wedge, L_{10}^\vee, L_{10}^\wedge, \ldots, L_{18}^\vee, L_{18}^\wedge\}$$

that define lattices. He found 32 such systems, each of them having 4 axioms.

In the same paper Petcu introduced 80 new axioms, which he called *absorptio-associative*:

$$((x \wedge y) \wedge z) \vee (((x \wedge y) \wedge z) \wedge t) = x \wedge (y \wedge z)$$

is a sample of these axioms. He looked for independent sets of axioms for lattices that can be constructed out of the 82-identity system Λ_5 which consists of the idempotent laws $L_1^\vee, L_1^\wedge$ and the 80 absorptio-associative axioms. He found 96 such systems, each of them having 3 axioms, namely two absorptio-associative laws and one idempotent law. Malliah [1971] introduced some more absorptio-associative laws and obtained 1240 new independent sets of axioms for lattices, among which 112 have 4 identities and 1128 consist of 3 identities. Petcu and Malliah also showed that many other combinations do not define lattices, but it is not known whether the systems they found exhaust all the minimal combinations that define lattices.

Felscher [1957], [1958] introduced eight new variants of the associative laws, among which

$L_{19}^{\vee}$ $\qquad x \vee (y \vee z) = (x \vee y) \vee (x \vee z)$,

$L_{19}^{\wedge}$ $\qquad x \wedge (y \wedge z) = (x \wedge y) \wedge (x \wedge z)$,

and used them in order to construct 11 new sets of identities defining lattices. One of these sets is the self-dual system

$$\mathbf{L}_2 = \{L_2^{\vee}, L_2^{\wedge}, L_4^{\vee}, L_4^{\wedge}, L_{19}^{\vee}, L_{19}^{\wedge}\} \ .$$

At this point we emphasize an idea which has been used many times in the axiomatics of lattices and Boolean algebras, as will be seen in this book. Namely, by changing the "normal" order of letters in certain axioms, one can prove commutativity, thus obtaining shorter systems of axioms.

So, for instance, Szász [1963] changed $L_{19}^{\vee}, L_{19}^{\wedge}$ to

$L_{20}^{\vee}$ $\qquad x \vee (y \vee z) = (y \vee x) \vee (z \vee x)$,

$L_{20}^{\wedge}$ $\qquad x \wedge (y \wedge z) = (y \wedge x) \wedge (z \wedge x)$,

and proved that the self-dual system

$$\mathbf{L}_3 = \{L_4^{\vee}, L_4^{\wedge}, L_{20}^{\vee}, L_{20}^{\wedge}\}$$

defines lattices. He also proved the independence of $\mathbf{L}_2, \mathbf{L}_3$ and of a system of axioms due to Klein-Barmen [1932]. Moreover, these proofs, as well as the proof of the independence of $\mathbf{L}_1$ given by Dubreil-Jacotin, Lesieur and Croisot [1953], are optimal, to the effect that the models used in them are of the shortest possible lengths.

Now we are going to prove that $\mathbf{L}_2$ and $\mathbf{L}_3$ actually define lattices. Since the commutativity $\mathbf{L}_2$ transforms axioms L_{19} into L_{20}, it suffices to do the proof for $\mathbf{L}_3$.

It was recalled before that axioms L_4 imply the idempotency L_1. From L_1 and L_{20} we infer commutativity:

$$x \vee y = x \vee (y \vee y) = (y \vee x) \vee (y \vee x) = y \vee x \ ,$$

and similarly for $L_2^{\wedge}$. Further we use L_4 to prove

$L_{21}^{\vee}$ $\qquad (x \vee y) \vee x = x \vee y$,

$L_{21}^{\wedge}$ $\qquad (x \wedge y) \wedge x = x \wedge y$.

Indeed,

$$(x \vee y) \vee x = (x \vee y) \vee ((x \vee y) \wedge x) = x \vee y$$

and similarly for $L_{21}^\wedge$. Finally we use L_{20}, L_{21} and commutativity to prove associativity, say $L_3^\vee$:

$$(x \vee y) \vee z = (x \vee z) \vee (y \vee z) = (x \vee (y \vee z)) \vee (z \vee (y \vee z))$$

$$= (x \vee (y \vee z)) \vee (y \vee z) = x \vee (y \vee z) .$$

$\square$

It is clear that, as was remarked above, several variants of system $\mathbf{L_2}$ can be constructed by replacing L_4 and/or L_{19} by variants of them such as L_5, L_6, L_7 and L_{20}, respectively; variants of L_8 and L_9 also arise naturally. Felscher (op.cit.) actually pointed out a few such sets of axioms, but it was Ruedin [1966], [1966/67], [1967], [1967/68], [1968] who undertook a laborious study in the same spirit as but apparently unaware of the work by Sorkin, Kalman, Rudeanu and Petcu. It should be mentioned that Ruedin related his research to the axiomatics of regular distributive groupoids (as he did for semilattices; cf.§1): a groupoid $(G, \cdot)$ is called *right* [*left*] *distributive* if it satisfies the identity

$$(x \cdot y) \cdot z = (x \cdot z) \cdot (y \cdot z) \ [\ x \cdot (y \cdot z) = (x \cdot y) \cdot (x \cdot z) \ ]$$

and *right* [*left*] *regular* provided the identity

$$x \cdot (y \cdot x) = y \cdot x \ [\ (x \cdot y) \cdot x = x \cdot y \ ]$$

holds. Ruedin found several independent sets of axioms for lattices, namely 44/24/3 sets having 6/5/4 axioms each. He also found many independent systems of axioms for lattices with least element 0 and for lattices with least element 0 and greatest element 1. It seems that Ruedin and Malliah are the last representatives of this kind of exhaustive research in lattice theory.

Birkhoff [1948], Problem 7 asked what are the the consequences of working with weakened forms of idempotency L_1 and absorption L_4, namely

L_{22} $x \wedge x = x \vee x$,

L_{23} $x \wedge (x \vee y) = x \vee (x \wedge y)$.

Answering this problem, Matusima [1952] devised 10 systems of 6–8 axioms for lattices, each of them containing the axioms L_2 and L_3 of commutativity and associativity, plus 2–4 other axioms chosen among L_{21}, L_{22}, L_{23} and several implications like $x \wedge y = x \vee y \implies x = y$, or $x \wedge y = y \wedge y \implies x \vee y = x$, etc. See also Padmanabhan [1971] and Appendix C.

McKenzie [1970] devised the self-dual system of axioms

$$\mathbf{L}_4 = \{L_{24}^\vee, L_{24}^\wedge, L_{25}^\vee, L_{25}^\wedge\} \, ,$$

where we have set

$L_{24}^\vee$ $x \vee (y \wedge (x \wedge z)) = x \, ,$

$L_{24}^\wedge$ $x \wedge (y \vee (x \vee z)) = x \, ,$

$L_{25}^\vee$ $((y \wedge x) \vee (x \wedge z)) \vee x = x \, ,$

$L_{25}^\wedge$ $((y \vee x) \wedge (x \vee z)) \wedge x = x \, ,$

and used it in order to obtain further a single identity defining lattices.

Let us prove that $\mathbf{L}_4$ actually characterizes lattices. We shall tacitly use the principle of duality. First we apply $L_{24}^\vee$ with $y := (y \vee x) \wedge (x \vee z)$ and $z := y \vee (x \vee z)$; taking into account $L_{24}^\wedge$ and $L_{25}^\wedge$, we obtain

$$x = x \vee (((y \vee x) \wedge (x \vee z)) \wedge (x \wedge (y \vee (x \vee z))))$$

$$= x \vee (((y \vee x) \wedge (x \vee z)) \wedge x) = x \vee x \, ,$$

showing that the two operations are idempotent. Hence if we take $y := x \vee z$ in $L_{24}^\wedge$ we obtain

$$x = x \wedge ((x \vee z) \vee (x \vee z)) = x \wedge (x \vee z) \, ,$$

therefore the two absorption laws L_4 hold.

Furthermore, taking $z := x$ in $L_{24}^\wedge$ we obtain the following variants of absorption:

$$x \wedge (y \vee x) = x \, , \ x \vee (y \wedge x) = x \, .$$

Therefore $L_{25}^\vee$ with $x := x \vee y$ and $z := x$ yields

$$x \vee y = ((y \wedge (x \vee y)) \vee (x \wedge x)) \vee (x \vee y) = (y \vee x) \vee (x \vee y) \, .$$

The latter equality implies

$$(y \vee x) \wedge (x \vee y) = (y \vee x) \wedge ((y \vee x) \vee (x \vee y)) = y \vee x$$

by using $L_4^\wedge$, and

$$y \vee x = (x \vee y) \vee (y \vee x)$$

by interchanging x and y. It follows from the last two identities and the second variant of absorption that

$$y \vee x = (x \vee y) \vee (y \vee x) = (x \vee y) \vee ((y \vee x) \wedge (x \vee y)) = x \vee y \, ,$$

showing that the two operations are commutative.

At this point we can borrow from lattice theory the following properties: the relation $\leq$ defined by $x \leq y \iff x \wedge y = x$ is reflexive, antisymmetric and satisfies $x \leq y \iff x \vee y = y$. Besides, if $y \leq x$ and $z \leq x$ then $y \vee z \leq x$ because $(y \vee z) \vee x = x$ by $\mathrm{L}_{25}^{\vee}$, where $y \wedge x = y$ and $x \wedge z = z$. On the other hand $x \leq x \vee (y \vee z)$ by $\mathrm{L}_4^{\wedge}$ with $y := y \vee z$, and $y \leq x \vee (y \vee z)$ by $\mathrm{L}_{24}^{\wedge}$ with x and y interchanged, hence $z \leq x \vee (z \vee y) = x \vee (y \vee z)$. Therefore

$$(x \vee y) \vee z \leq x \vee (y \vee z) = (z \vee y) \vee x \leq z \vee (y \vee x) = (x \vee y) \vee z \,,$$

which proves the associativity of the two operations.   □

Two other characterizations of lattices generalize Theorem 1.1.

Theorem 1.2.1. (Petcu [1971]) *An algebra $(L, \vee, \wedge)$ of type $(2,2)$ is a lattice if and only if it satisfies the following two conditions, for every n and every $n+1$ variables $x_1, \ldots, x_n, y$:*

(i) *every two $\vee-$expressions φ and ψ in the variables $x_1, \ldots, x_n$ generate the identity $\varphi \wedge (\varphi \vee y) = \psi$;*

(ii) *every two $\wedge-$expressions φ and ψ in the variables $x_1, \ldots, x_n$ generate the identity $\varphi \vee (\varphi \wedge y) = \psi$.*

PROOF: If L is a lattice then the identity $\varphi \wedge (\varphi \vee y) = \varphi$ holds. On the other hand, if φ, ψ are $\vee-$expressions in the variables $x_1, \ldots, x_n$, then they generate the same function by Theorem 1.1. The proof of (ii) is similar.

Conversely, suppose conditions (i) and (ii) are satisfied. Then, taking $\varphi := \psi := x$, we obtain the absorption laws. Therefore conditions (i) and (ii) reduce to the following: every two $\vee-$expressions $(\wedge-$expressions$)$ φ and ψ in the same variables generate the same function. In view of Theorem 1.1, this implies that $(L, \vee)$ and $(L, \wedge)$ are semilattices.   □

Theorem 1.2.2. (Petcu [1971]) *An algebra $(L, \vee, \wedge)$ of type $(2,2)$ is a lattice if and only if it satisfies the following two conditions, for every n and every $n+2$ variables $x_1, \ldots, x_n, y, z$:*

(j) *if φ and ψ are $\vee-$ expressions or $\wedge-$expressions in the variables $x_1, \ldots, x_n$, then they generate the identity*

$$\varphi \wedge (\varphi \vee y) = \psi \vee (\psi \wedge z) \,;$$

(jj) *the identity $\varphi(x, \ldots, x) = x$ holds.*

PROOF: If L is a lattice then (j) and (jj) follow from Theorem 2.1 and Lemma 1.2, respectively.

Conversely, suppose conditions (j) and (jj) are satisfied. Taking $\varphi :=$ $\psi := x := z$ in (j) and applying (jj) twice we obtain

$$x \wedge (x \vee y) = x \vee (x \wedge x) = x \vee x = x$$

and $x \vee (x \wedge y) = x$ is obtained similarly. Now condition (j) reduces to conditions (i) and (ii) in Theorem 2.1, therefore L is a lattice.   □

To state the following results we need to generalize the prerequisites used in §1. Given an algebra $(L, \vee, \wedge)$ of type (2,2), the set of $\vee, \wedge-terms$ is the least set of words obeying the rules 1) every variable is a term, and 2) if φ and ψ are terms, then $(\varphi \vee \psi)$ and $(\varphi \wedge \psi)$ are terms. The functions generated by $\vee, \wedge-terms$ are said to be $\vee, \wedge-polynomials$; if L is a lattice, they are also called *lattice polynomials*. Furthermore, identities are defined as in §1.

Lemma 1.2.1. *If the algebra* $(L, \vee, \wedge)$ *of type (2,2) satisfies* $\mathrm{L}_1^\vee$ *and* $\mathrm{L}_1^\wedge$, *then every* $\vee, \wedge-polynomial$ f *satisfies* $f(x, \ldots, x) = x$.

PROOF: Similar to the proof of Lemma 1.2.   □

Corollary 1.2.1. *Every lattice polynomial* f *satisfies* $f(x, \ldots, x) = x$.

A class of lattices is called *finitely definable* if it is defined within the class of all lattices by a finite set of identities, that is, if the class consists of those lattices that satisfy a certain finite set of identities.

Theorem 1.2.3. (Padmanabhan [1968]) *Every finitely definable class of lattices can be defined by a single identity within the class of all lattices.*

PROOF: It suffices to show that any set of two identities

(3) $f_1 = g_1 \ \& \ f_2 = g_2$

is equivalent to the single identity

(4) $f_1(x_1, \ldots, x_n) \vee f_2(y_1, \ldots, y_p) = g_1(x_1, \ldots, x_n) \vee g_2(y_1, \ldots, y_p) ,$

where the variable sets $\{x_1, \ldots, x_n\}$ and $\{y_1, \ldots, y_p\}$ are disjoint.

Clearly (3) implies (4). Conversely, suppose identity (4) holds. Taking $y_1 := \cdots := y_p := g_1(x_1, \ldots, x_n)$ and using Lemma 1.1 we get

$$f_1(x_1, \ldots, x_n) \vee g_1(x_1, \ldots, x_n) = g_1(x_1, \ldots, x_n) \vee g_1(x_1, \ldots, x_n) = g_1(x_1, \ldots, x_n).$$

Taking $y_1 := \cdots := y_p := f_1(x_1, \ldots, x_n)$ we obtain

$$f_1(x_1, \ldots, x_n) = f_1(x_1, \ldots, x_n) \vee f_1(x_1, \ldots, x_n) = g_1(x_1, \ldots, x_n) \vee f_1(x_1, \ldots, x_n),$$

hence $f_1(x_1,\ldots,x_n) = g_1(x_1,\ldots,x_n)$ and similarly we prove that $f_2(y_1,\ldots,y_p) = g_2(y_1,\ldots,y_p)$. ☐

Another major direction of research consists in looking for systems with as few identities as possible. It seems that for lattices this trend was inaugurated by Sorkin and Ponticopoulos, independently of each other. Sorkin [1962] defined lattices by the two absorption laws L_4 plus a third identity with 23 occurrences of 9 variables, while Ponticopoulos [1962] used the idempotency laws L_1 plus a third identity with 16 occurrences of 9 variables. We have already referred to the numerous three-identity systems due to Petcu and Malliah. Another such system is

$$\mathbf{L_5} = \{L_{24}^\wedge, L_{25}^\vee, L_{26}^\wedge\}\,,$$

given by McCune and Padmanabhan [1996], where we have set

$L_{26}^\wedge$ $((x \vee y) \wedge (x \vee z)) \wedge x = x\,.$

A class of algebras is said to be *n-based* if it can be defined by a set of n identities, also known as a *basis* of the class. Sorkin [1962] proved that any class of lattices defined by finitely many identities is 3–based. This result was improved in the case of lattices.

Theorem 1.2.4. (Padmanabhan [1968]) *Let f and g be $\vee, \wedge-$polynomials of n variables $x_1,\ldots,x_n$ over an algebra $(L,\vee,\wedge)$ of type $(2,2)$. Then L is a lattice satisfying the identity $f = g$ if and only if it fulfils the following identities:*

$L_{27}^\vee$ $(x \wedge y) \vee y = y$ (which is the same axiom as $L_7^\vee$),

$L_{28}^\vee$ $(((x \wedge f) \wedge z) \vee u) \vee v = (((g \wedge z) \wedge x) \vee v) \vee ((t \vee u) \wedge u)\,.$

PROOF: Put $z := u$, $x := v$ in $L_{28}^\vee$; by $L_{27}^\vee$ we get

(5) $u \vee v = v \vee ((t \vee u) \wedge u)\,.$

Putting $t := x \wedge u$ in the above and using $L_{27}^\vee$ we have

(6) $u \vee v = v \vee (u \wedge u)\,.$

Putting $u := x \wedge v$ in the above and applying $L_{27}^\vee$ we get

(7) $v = v \vee ((x \wedge v) \wedge (x \wedge v))\,.$

Putting $v := (u \wedge u) \wedge (u \wedge u)$ in (6) and using (7) and $L_{27}^\vee$ we get

(8) $$u = u \wedge u \ .$$

Put $x :=: y =: u$ in $\mathrm{L}_{27}^{\vee}$; by (8) we have

(9) $$u \vee u = u \ .$$

Put $v := u$ in (5); by (9) we get

(10) $$u = u \vee ((t \vee u) \wedge u) \ .$$

Put $v := (t \vee u) \wedge u$ in (5); again by (9) we have

(11) $$u \vee ((t \vee u) \wedge u) = (t \vee u) \wedge u \ .$$

From (10) and (11) we see that

(12) $$u = (t \vee u) \wedge u \ .$$

So, (5) reads as

(13) $$u \vee v = v \vee u \ .$$

Now $\mathrm{L}_{28}^{\vee}$ becomes

(14) $$(((x \wedge f) \wedge z) \vee u) \vee v = (((g \wedge z) \wedge x) \vee v) \vee u \ .$$

Put $x := x_1 = \cdots := x_n := z$. It follows from (8) and (9) via Lemma 2.1 that

$$(x \wedge f(x, \ldots, x)) \wedge x = x = (g(x, \ldots, x) \wedge x) \wedge x \ ,$$

hence (14) reduces to

(15) $$(x \vee u) \vee v = (x \vee v) \vee u \ .$$

Therefore, by (15) and (13),

$$(x \vee u) \vee v = (v \vee x) \vee u = (v \vee u) \vee x \ ;$$

by applying (13) twice, we get

(16) $$(x \vee u) \vee v = x \vee (u \vee v) \ .$$

Thus $(L, \vee)$ is a semilattice.

Now we take $u := v$ in (14) and obtain

$$((x \wedge f) \wedge z) \vee u = ((g \wedge z) \wedge x) \vee u \ ,$$

which implies, by taking in turn $u := (x \wedge f) \wedge z$ and $u := (g \wedge z) \wedge x$, that

(17) $$(x \wedge f) \wedge z = (g \wedge z) \wedge x \ .$$

Further we take $x_1 := \cdots := x_n := y$ in (17) and using again Lemma 2.1 we obtain $(x \wedge y) \wedge z = (y \wedge z) \wedge x$. The latter identity and (8) show that $(L, \wedge)$ is a semilattice according to system $\{S_1, S_6\}$ in §1. Since the absorption laws L_{27} and (12) also hold, it follows that $(L, \vee, \wedge)$ is a lattice.

Finally we obtain the identity $f = g$ by taking $x := z := f \vee g$ in (17).
□

Note that equation L_{28} involves $n + 5$ variables, while the letters f, g, x, z, u, v and t have 12 ocurrences. The paper by Padmanabhan [1969b] establishes a result which resembles Theorem 2.4, except that instead of L_{25} there is a similar equation with 14 occurrences of letters.

Corollary 1.2.2. (Kalman [1968]) *Lattices are characterized by the identities* $L_{27}^{\vee}$ *and*

$L_{29}^{\vee}$ $(((x \wedge y) \wedge z) \vee u) \vee v = (((y \wedge z) \wedge x) \vee v) \vee ((t \vee u) \wedge u)$.

PROOF: Characterize the class of all lattices by the identity $y = y$.   □

Corollary 1.2.3. *Every finitely definable class of lattices can be characterized by two identities, namely* L_{27} *and an identity of the form* L_{28} .

PROOF: By Theorems 2.3 and 2.4.   □

Tamura [1975] devised the system

$$\mathbf{L}_6 = \{L_{27}^{\vee}, L_{30}^{\vee}\} \, ,$$

where $L_{30}^{\vee}$ is a slight improvement of $L_{29}^{\vee}$:

$L_{30}^{\vee}$ $(((x \wedge y) \wedge z) \vee u) \vee v = (((y \wedge z) \wedge x) \vee v) \vee ((y \vee u) \wedge u)$.

Padmanabhan [1972] suggested the system

$$\mathbf{L}_7 = \{L_{31}^{\wedge}, L_{32}^{\vee}\} \, ,$$

where we have set

$L_{31}^{\wedge}$ $(x \vee y) \wedge z = (z \wedge (x \vee x)) \vee (z \wedge (y \vee x))$,

$L_{32}^{\vee}$ $((z \wedge x) \vee (z \wedge y)) \vee ((z \wedge z) \vee (z \wedge z)) = z$.

A self-dual system of identities defining lattices was given by Padmanabhan [1983], namely

$$\mathbf{L}_8 = \{L_7^{\vee}, L_7^{\wedge}, L_{33}^{\vee}, L_{33}^{\wedge}\} \, ,$$

where

$L_{33}^{\vee}$ $(x \vee y) \vee z = (y \vee z) \vee x$,

$L_{33}^{\wedge}$ $(x \wedge y) \wedge z = (y \wedge z) \wedge x$, while in fact $L_7^\vee$ and $L_7^\wedge$ can be replaced by any other variants of the absorption laws. The proof is very easy: the absorption laws imply the idempotency laws (cf. $\mathbf{L_1}$), therefore $\vee$ and $\wedge$ are semilattice operations according to Padmanabhan's system $\{S_1, S_6\}$.

Tamura [1975] characterized lattices with 0 by the identities $L_{27}^\vee$ and $L_{34}^\vee$, where

$L_{34}^\vee$ $\qquad (((((0 \vee a) \wedge b) \wedge c) \vee d) \vee e = (((b \wedge c) \wedge a) \vee e) \vee ((b \vee d) \wedge d)$,

while the following system of two identities for the lattices with 0 and 1 is due to Sobociński [1979]:

$L_{35}^\vee$ $\qquad (y \wedge (z \wedge x)) \vee x = x$,

$L_{36}^\vee$ $\qquad ((x \wedge (y \wedge z)) \vee t) \vee u = ((((z \wedge 1) \vee 0) \wedge (x \wedge y)) \vee u) \vee ((v \vee t) \wedge t)$.

Quite recently, McCune and Padmanabhan found a new system of axioms:

Theorem 1.2.5. *The following self-dual set of two identities characterizes lattices:*

$L_{62}^\vee$ $\qquad (((x \wedge y) \vee y) \wedge (z \vee y)) \vee (u \wedge ((v \wedge y) \vee (y \wedge w))) = y$,

$L_{62}^\wedge$ $\qquad (((x \vee y) \wedge y) \vee (z \wedge y)) \wedge (u \vee ((v \vee y) \wedge (y \vee w))) = y$.

See Appendix A for a proof provided by the computer program Prover9.

1.3. One-Based Theories

From the point of view of this section, the *equational theory* of a class of algebras is the collection of all *equations* (i.e. identities) that hold in all members of that class. As a trivial example, the equational theory of a class of one-element algebras consists of all identities of the relevant type and is generated by the single identity $x = y$. In the other extreme of the spectrum, the equational theory of the class of all structures of a given type contains only equations of the form $x = x$. For more details the reader is referred to the excellent survey article on this topic by Tarski [1968].

Recall that a class of algebras is said to be n-based if it can be defined by a set of n identities. Then that class is a variety and we will alternatively say that the equational theory of that variety is n-based.

Given a finitely-based equational theory T of algebras, it is but natural to ask for the minimum number of equations that a basis for T can contain, and in particular, to determine whether T has a basis consisting of a single identity. The answer is well known for group-like systems. Every finitely-based theory of groups (or loops) is always one-based (results due to Higman and Neumann [1952], Tarski [1968], Padmanabhan [1969a]). These algebras admit cancellation laws and have quasi-group properties, and these play a crucial role in constructing a single axiom for group-like theories. However, lattices neither admit any cancellation laws nor they enjoy any meaningful quasi-group property. In this sense, the equational theory of lattices is "far removed" from that of group theory. In view of these intuitive observations, it was widely believed - till 1967 - that the equational theory of lattices may not be one-based and, in fact, not even definable by *any set* of identities of the form $f(x, x_1, \ldots, x_n) = x$, which is, prima facie, essential for any potential one-based theory. It was McKenzie [1970] who first published the theorem that the variety of all lattices can, indeed, be defined by such "absorption laws" and that the equational theory is, in fact, one-based. McKenzie also mentioned in the same publication that no other variety of lattices (except the variety of singletons, defined by $x = y$) is one-based. Hence for the finitely based lattice varieties "two" (as proved in Theorem 2.4) is best possible.

In this section we regard semilattices as algebras of type (2) and lattices as algebras of type (2,2) (disjunction and conjunction). We will show that while semilattices cannot be defined by a single axiom, lattices can be so defined.

We need to introduce some terminlogy. By an *absorption identity* we mean an identity of the form $f(x, x_1, \ldots, x_n) = x$. An identity $f = g$ is called *regular* if the sets of variables occurring on the two sides of the equation are the same.

Lemma 1.3.1. *Any system of identities defining a class of lattices contains an absorption identity.*

PROOF: A non-absorption identity is of the form $f = g$, where neither f nor g is a variable and hence both f and g contain at least one of the operation symbols $\vee, \wedge$. Now take the two-element set $\{0, 1\}$ and define $x \vee y = x \wedge y = 0$ for all x and y. This algebra will satisfy all the non-absorption identities but $1 \vee 1 = 0 \neq 1$ and hence is not idempotent.   □

Lemma 1.3.2. *Any identity valid in a semilattice is regular.*

PROOF: The axioms S_1, S_2, S_3 are regular and regularity is preserved under equational consequences (by substitutions).   □

Theorem 1.3.1. (Potts [1965]) *The variety of all semilattices cannot be defined by a single identity.*

PROOF: It follows from Theorems 3.1 and 3.2 that a potential single identity for the class of all semilattices must be of the form $f(x, \dots, x) = x$. Suppose x occurs $n+1$ times on the left-hand side. If $n > 1$ take the additive group $(\mathbf{Z}_n, +)$ and define $x \vee y = x + y$. Then $f(x, \dots, x) = (n+1)x = x$, but this model of the identity $f = x$ is not idempotent. If $n = 1$ the equation is $x \circ x = x$ and it is very easy to produce a three-element groupoid which is idempotent but not associative. Contradiction. □

The next theorem uses the concept of *Jónsson term*, also called *majority polynomial*. This means a ternary polynomial p satisfying the identities

(1) $$p(x, x, z) = p(x, y, x) = p(y, x, x) = x .$$

For instance, lattices admit the majority polynomial

$$p(x, y, z) = (x \wedge y) \vee (y \wedge z) \vee (x \wedge z) .$$

Theorem 3.2 below provides a very simple identity to show that any finitely-based variety of algebras that admits a Jónsson term and is definable by absorption identities, is one-based. Our knowledge of this result is due to the paper by McKenzie [1970] and forms a part of Theorem 1.2 stated there without proof.

Lemma 1.3.3. (Padmanabhan [1977]) *Let T be an equational theory with a majority polynomial p. For arbitrary polynomials f and g, the validity of two identities $f = x$ and $g = x$ in T is equivalent to $p(f, g, y) = x$, where y is a variable not occurring in f or g.*

PROOF: Clearly $f = x$ and $g = x$ together imply $p(f, g, y) = x$. To get the converse, substitute $y := f$ to derive $f = x$ and $y := g$ to derive $g = x$. □

Theorem 1.3.2. (Padmanabhan [1977]) *Let T be an equational theory which is defined by finitely many absorption identities and has a majority polynomial. Then T is one-based.*

PROOF: By Lemma 3.3 we can assume that T has a basis of the form $f = y$ and the three identities (1) which define a majority polynomial p. Now consider

(2) $$p(p(x, y, y), u, p(p(x, y, y), f, z)) = y ,$$

where z and u are variables not occurring in f. Certainly T implies the identity (2). Conversely, let us prove that (2) implies (1) and $f = y$.

Put $z := p(p(x, y, y), f, w)$ in (2). Since $p(p(x, y, y), f, p(p(x, y, y), f, w)) = y$ by (2), we obtain

$$(3) \qquad\qquad p(p(x, y, y), u, y) = y \ .$$

Putting $x := p(a, y, y)$ in (3) and noting that $p(p(a, y, y), y, y) = y$ again by (3), we get

$$(4) \qquad\qquad y = p(p(p(a, y, y), y, y), u, y) = p(y, u, y) \ .$$

Put $z := p(x, y, y)$ in (2). By two successive applications of (4) we have

$$(5) \qquad \begin{aligned} y &= p(p(x, y, y), u, p(p(x, y, y), f, p(x, y, y))) \\ &= p(p(x, y, y), u, p(x, y, y)) = p(x, y, y) \ . \end{aligned}$$

Thus (2) becomes $p(y, u, p(y, f, z)) = y$, which, by substituting $u := p(y, f, z)$ and using (5), yields

$$(6) \qquad\qquad y = p(y, p(y, f, z), p(y, f, z)) = p(y, f, z) \ .$$

Finally, $z := f$ in (6) implies, by (5) again, the identity

$$(7) \qquad\qquad\qquad f = y \ ,$$

which, in turn, reduces (6) to

$$(8) \qquad\qquad\qquad p(y, y, z) = y \ .$$

Thus p is a majority polynomial by (4), (5) and (8), and moreover, we have the identity $f = y$. $\qquad\qquad\qquad\qquad\qquad\qquad\qquad\qquad\square$

As was mentioned in the beginning, McKenzie first constructed a single-equation basis for lattices, involving 34 variables. However, starting from the four absorption identities given by McKenzie [1970] ($\lambda_1, \lambda_2, \lambda_3, \lambda_4$ on page 27) and using Lemma 3.3 we get an identity $f = y$ with five variables, and applying Theorem 3.3 to this identity and the three identities (1), we obtain a single axiom for lattices in only seven variables. More generally, if $\mathbf{K}$ is an equational class defined by n absorption identities involving at most k variables and if $\mathbf{K}$ admits a majority polynomial, then $\mathbf{K}$ has a one-basis involving at most $k + 2 + \log_2 n$ variables (G.M. Bergamn, Reno

AMS Conference). On the other hand, Grätzer [1998] (cf. Problem 17) asks for a short identity defining lattices. The following table reports numerical characteristics of several single axioms defining lattices:

Reference	Variables	Length
McKenzie [1970]	34	300,000
Padmanabhan [1977]	7	243
McCune and Padmanabhan [1996a,b]	7	79
Veroff [2001]	8	77
McCune, Padmanabhan and Veroff [2003]	8	29

Thus the latter paper provides the shortest known single axiom for the equational theory of all lattices, having length 29 and 8 variables:

$L_{36}^{\wedge}$ $(((y\vee x)\wedge x)\vee(((z\wedge(x\vee x))\vee(u\wedge x))\vee v))\wedge(w\vee((s\vee x)\wedge(x\vee t)))) = x$,

the length being understood as the number of occurrences of variables and operation symbols.

This single-identity for lattices was found with the aid of a computer program called Otter (Organized Tools and Techniques for Efficient Research), devised by McCune; cf. McCune and Padmanabhan [1996a]. The search strategy was the following:

– generate candidates of the form $f = x$;

– eliminate candidates that are not lattices identities by incorporating certain equational filters in the program;

– for each candidate, try either to find a small finite nonlattice model of it, or to derive from it the standard system L_1.

Let us explain the elimination of candidates by an actual example encountered by the machine in the process of discovering a single axiom for lattices. Otter discovered the identity

(L^*) $(y \vee ((x \vee z) \wedge (z \vee x))) \wedge (((x \wedge t) \vee x) \vee ((u \wedge x) \vee (x \wedge w))) = x$

and it is easy to verify that it is valid in all lattices. However it is not strong enough to derive all the axioms for lattices. The reason is that we can "parse" the above identity and find a stronger class of lattice identities which is well known to be inadequate for defining lattices. Here is the parsing process:

$$(x \wedge t) \vee x = x ,$$

$$x \vee ((u \wedge x) \vee (x \wedge w)) = x ,$$

$$(y \vee ((x \vee z) \wedge (z \vee x))) \wedge x = x \,.$$

Although these three identities do imply the single identity (L*), there is a non-lattice model satisfying them, hence (L*) as well, therefore (L*) is not a single identity for lattice theory. Here is one such model.

Take the five-element lattice $\{0, c, a, b, 1\}$ with $0 < c = a \wedge b < a \vee b = 1$. Re-define $a \wedge b = b \wedge a = 0$, otherwise let $\vee$ and $\wedge$ be the lattice operations. Since the join operation is a semilattice operation, the relation $x \leq y$ defined by $x \vee y = y$ is a partial order. Also, this algebra satisfies all the two-variable lattice laws, because the subalgebra generated by any two elements is, indeed, a lattice. Therefore

$$(x \wedge t) \vee x = x \,,$$

$$x \vee ((u \wedge x) \vee (x \wedge w)) = ((x \vee (u \wedge x)) \vee (x \wedge w) = x \vee (x \wedge w) = x \,,$$

$$(y \vee ((x \vee z) \wedge (z \vee x))) \wedge x = (y \vee (x \vee z)) \wedge x = x \,.$$

However $(a \wedge b) \wedge c = 0 \wedge c = 0$, while $a \wedge (b \wedge c) = a \wedge c = c$. Thus identity (L*) is eliminated. Such counter-examples are incorporated in the software, so that it can filter out the non-lattice identities automatically.

The program was run on several hundred processors, usualy in jobs of 10-20 hours, over a period of several weeks. Two short single-identities for lattices were found, namely $\mathbf{L}_{36}^{\wedge}$ and

$\mathbf{L}_{37}^{\wedge}$ $(((y \vee x) \wedge x) \vee (((z \wedge (x \vee x)) \vee (u \wedge x)) \wedge v)) \wedge (((w \vee x) \wedge (s \vee x)) \vee t) = x \,,$

while for many shorter equations and equations with fewer variables, neither a proof of $\mathbf{L}_1$ nor a nonlattice model could be found (caution: only a nonlattice model of an equation eliminates it from the list of candidates!).

The program consists in fact of several programs with specialized jobs. The program for proof searching, called Otter, derived $\mathbf{L}_1$ from $\mathbf{L}_{36}^{\wedge}$ in more than 250 steps. Then Otter was used to obtain $\mathbf{L}_4$, which it did in about 170 steps. Later on, L. Wos used various methods to simplify the Otter proof and obtained a proof in 50 steps. Each step uses paramodulation, an inference rule that combines variable instantiation (or unification) and equality substitution into one step.

We give below a variant of the latter proof. We indicate paramodulation in the following form: "take <substitution S1> in i, then use j with <substitution S2>". More exactly, let $i_1 = i_2$ and $j_1 = j_2$ be the equations i and j, respectively. Let $i'_1 = i'_2$ and $j'_1 = j'_2$ be the equations S1(i) and S2(j), respectively. The resulting equation is $i''_1 = i'_2$, where i''_1 is obtained from i_1' by replacing the subterm(s) j_1' by j_2' (or $i_1'' = i_2''$, where i_2'' is obtained from i_2' by the same transformation). For instance, "Take $y := x \wedge y$ in $x \wedge (x \vee y) = x$, then use $x \vee (x \wedge y) = x$" produces the equation $x \wedge x = x$; here S2 is the identity.

1. $(((y \vee x) \wedge x) \vee (((z \wedge (x \vee x)) \vee (u \wedge x)) \wedge v)) \wedge (w \vee ((s \vee x) \wedge (x \vee t))) = x$.
This is axiom $\mathrm{L}_{36}^{\wedge}$.

2. $(((x \vee y) \wedge y) \vee (y \vee y)) \wedge (z \vee ((u \vee y) \wedge (y \vee v))) = y$.
Set $y \vee y = Y$. Take $x := y$, $y := x$, $z := y \vee Y$, $u := (z \wedge (Y \vee Y)) \vee (u \wedge Y)$, $v := w \vee ((s \vee Y) \wedge (Y \vee t))$, $w := z$, $s := u$, $t := v$ in 1, then use 1 with $x := Y$, $v := y$.

3. $(((x \vee (y \vee y)) \wedge (y \vee y)) \vee ((y \vee y) \vee (y \vee y))) \wedge (z \vee y) = y \vee y$.
Take $y := y \vee y$, $u := (x \vee y) \wedge y$, $v := (u \vee y) \wedge (y \vee v)$ in 2, then use 2 with $z := y \vee y$.

4. $(((x \vee y) \wedge y) \vee (((y \vee y) \vee (z \wedge y)) \wedge u)) \wedge (v \vee ((w \vee y) \wedge (y \vee t))) = y$.
Take $x := y$, $y := x$, $z := ((x \vee (y \vee y)) \wedge (y \vee y)) \vee ((y \vee y) \vee (y \vee y))$, $u := z$, $v := u$, $w := v$, $s := w$ in 1, then use 3 with $z := y$.

5. $(((x \vee Y) \wedge Y) \vee (Y \vee Y)) \wedge (v \vee y) = Y$, where $Y = ((y \vee y) \vee (z \wedge y)) \wedge u$.
Take $y := Y$, $z := v$, $u := (x \vee y) \wedge y$, $v := (w \vee y) \wedge (y \vee t)$ in 2, then use 4 with $v := Y$.

6. $(((x \vee y) \wedge y) \vee (((((y \vee y) \vee (z \wedge y)) \wedge u) \vee (v \wedge y)) \wedge w)) \wedge (t \vee ((s \vee y) \wedge (y \vee r))) = y$.
Set $((y \vee y) \vee (z \wedge y)) \wedge u = Y$. Take $x := y$, $y := x$, $z := ((x \vee Y) \wedge Y) \vee (Y \vee Y)$, $u := v$, $v := w$, $w := t$, $t := r$ in 1, then use 5 with $v := y$.

7. $(((x \vee y) \wedge y) \vee (z \wedge y)) \wedge (u \vee ((v \vee y) \wedge (y \vee w))) = y$.
Set $z \wedge y = Y$. Take $u := Y$, $v := (((Y \vee Y) \vee (z \wedge Y)) \wedge u) \vee (v \wedge Y)$, $w := t \vee ((s \vee Y) \wedge (Y \vee r))$, $t := u$, $s := v$, $r := w$ in 6, then use 6 with $x := y \vee y$, $y := Y$, $w := y$.

8. $(((x \vee (y \wedge z)) \wedge (y \wedge z)) \vee (u \wedge (y \wedge z))) \wedge (v \vee z) = y \wedge z$.
Take $y := y \wedge z$, $z := u$, $u := v$, $v := (x \vee z) \wedge z$, $w := (v \vee z) \wedge (z \vee w)$ in 7, then use 7 with $y := z$, $z := y$, $u := y \wedge z$.

9. $(((x \vee y) \wedge y) \vee (((z \wedge y) \vee (u \wedge y)) \wedge v)) \wedge (w \vee ((t \vee y) \wedge (y \vee s))) = y$.
Take $x := y$, $y := x$, $z := ((x \vee (z \wedge y)) \wedge (z \wedge y)) \vee (u \wedge (z \wedge y))$, $s := t$, $t := s$ in 1, then use 8 with $y := z$, $z := y$, $v := y$.

10. $(((x \vee y) \wedge y) \vee y) \wedge (z \vee ((u \vee y) \wedge (y \vee v))) = y$.
Take $z := x \vee y$, $u := (z \wedge y) \vee (u \wedge y)$, $v := w \vee ((t \vee y) \wedge (y \vee s))$, $w := z$, $t := u$, $s := v$ in 9, then use 9 with $v := y$.

11. $(((x \vee y) \wedge y) \vee (((z \wedge y) \vee (u \wedge y)) \wedge v)) \wedge (w \vee y) = y$.
Take $t := (x \vee y) \wedge y$, $s := (u \vee y) \wedge (y \vee v)$ in 9, then use 10 with $z := y$.

12. $(((x \vee y) \wedge y) \vee (z \wedge y)) \wedge (u \vee y) = y$.

Take $v := (x \vee y) \wedge y$, $w := (x \vee y) \wedge (y \vee v)$ in 7, then use 10 with $z := y$, $u := x$.

13. $(((x \vee y) \wedge y) \vee (y \vee y)) \wedge (z \vee y) = y$.

Take $u := (x \vee y) \wedge y$, $v := (u \vee y) \wedge (y \vee v)$ in 2, then use 10 with $z := y$, $u := x$.

14. $(x \vee (y \wedge (x \vee x))) \wedge (z \vee ((u \vee (x \vee x)) \wedge ((x \vee x) \vee v))) = x \vee x$.

Take $x := (x \vee x) \wedge x$, $y := x \vee x$, $z := y$, $u := z$, $v := u$, $w := v$ in 7, then use 13 with $y := x$, $z := x$.

15. $(x \vee x) \wedge (y \vee ((z \vee (x \vee x)) \wedge ((x \vee x) \vee u))) = x \vee x$.

Take $y := ((x \vee x) \wedge x) \vee (x \vee x)$, $z := y$, $u := z$, $v := u$ in 14, then use 13 with $y := x$, $z := x$.

16. $(((x \vee y) \wedge y) \vee ((z \wedge y) \vee (z \wedge y))) \wedge (u \vee y) = y$.

Set $X = z \wedge y$. Take $u := z$, $v := y \vee ((z \vee (X \vee X)) \wedge ((X \vee x) \vee u))$, $w := u$, in 11, then use 15 with $x := X$, $w := u$.

17. $((x \wedge y) \vee (x \wedge y)) \wedge (z \vee y) = (x \wedge y) \vee (x \wedge y)$.

Take $x := x \wedge y$, $y := z$, $z := (x \vee y) \wedge y$, $u := y$ in 15, then use 16 with $z := x$, $u := (x \wedge y) \vee (x \wedge y)$.

18. $((x \vee y) \wedge y) \vee ((x \vee y) \wedge y) = y$.

Take $x := x \vee y$ in 17, then use 12 with $z := x \vee y$, $u := z$.

19. $(x \wedge y) \wedge (z \vee y) = x \wedge y$.

Take $y := x$, $z := y$, $u := x \vee (x \wedge y)$, $v := z$ in 8, then use 18 with $y := x \wedge y$, $z := x$.

20. $x \wedge (y \vee x) = x$.

Take $x := y$, $y := x$, $z := y \vee x$, $u := y$ in 12, then use 18 with $x := y$, $y := x$.

21. $x \wedge (y \vee ((z \vee x) \wedge (x \vee u))) = x$.

Take $x := y$, $y := x$, $z := y \vee x$, $u := y$, $v := z$, $w := u$ in 7, then use 18 with $x := y$, $y := x$, $z := y$.

22. $(((x \vee y) \wedge y) \vee ((z \wedge y) \vee (u \wedge y))) \wedge (v \vee y) = y$.

Take $v := x \vee ((z \wedge y) \vee (u \wedge y))$, $w := v$ in 11, then use 20 with $x := (z \wedge y) \vee (u \wedge y)$, $y := x$.

23. $((x \wedge y) \vee (z \wedge y)) \wedge (u \vee y) = (x \wedge y) \vee (z \wedge y)$.

Take $x := (x \wedge y) \vee (z \wedge y)$, $y := u$, $z := (x \vee y) \wedge y$, $u := y$ in 21, then use 22 with $z := x$, $u := z$, $v := (x \wedge y) \vee (z \wedge y)$.

24. $((x \vee y) \wedge y) \vee (z \wedge y) = y$.

Take $x := x \vee y$ in 23, then use 12.

25. $x \wedge (x \vee y) = x$.

Take $y := (x \vee (x \vee y)) \wedge (x \vee y)$, $u := y$ in 21, then use 24 with $y :=$

$x \vee y, \; z = z \vee x.$

26. $((x \vee y) \wedge y) \vee ((z \wedge y) \vee (u \wedge y)) = y.$

Take $v := ((x \vee y) \wedge y) \vee ((z \wedge y) \vee (u \wedge y))$ in 22, then use 25 with $x := ((x \vee y) \wedge y) \vee ((z \wedge y) \vee (u \wedge y)).$

27. $((x \vee y) \wedge y) \vee y = y.$

Take $z := ((x \vee y) \wedge y) \vee y$ in 10, then use 25 with $x := ((x \vee y) \wedge y) \vee y$, $y := (u \vee y) \wedge (y \vee v).$

28. $x \wedge (y \vee (x \vee z)) = x.$

Take $y := x \vee z, \; z := y$ in 19, then use twice 25 with $y := z.$

29. $(x \wedge x) \vee x = x.$

Take $x := (x \vee x) \wedge x, \; y := x$ in 27, then use 27 with $y := x.$

30. $x \wedge (y \vee (x \wedge (x \vee z))) = x.$

Take $z := (y \vee x) \wedge x, \; u := z$ in 21, then use 27 with $x := y, \; y := x.$

31. $x \wedge x = x.$

Take $y := (y \vee x) \wedge x$ in 20, then use 27 with $x := y, \; y := x.$

32. $x \vee x = x.$

By 29 via 31.

33. $x \wedge ((y \vee x) \wedge (x \vee z)) = x.$

Take $y := (y \vee x) \wedge (x \vee z), \; z := y, \; u := z$ in 21, then use 32 with $x := (y \vee x) \wedge (x \vee z).$

34. $(x \vee y) \wedge y = y.$

Take $z := x \vee y$ in 24, then use 32 with $x := (x \vee y) \wedge y.$

35. $x \wedge (y \wedge x) = y \wedge x.$

Take $x := (x \vee x) \wedge x, \; y := y \wedge x$ in 34, then use 24 with $y := x, \; z := y.$

36. $(x \vee (((y \wedge x) \vee (z \wedge x)) \wedge u)) \wedge (v \vee x) = x.$

Take $x := y, \; y := x, \; z := y, \; u := z, \; v := u, \; w := v$ in 11, then use 34 with $x := y, \; y := x.$

37. $((x \vee y) \wedge (y \vee z)) \wedge y = y \wedge ((x \vee y) \wedge (y \vee z)).$

Take $x := (x \vee y) \wedge (y \vee z)$ in 35, then use 33 with $x := y, \; y := x, \; z := x.$

38. $(x \vee (x \wedge y)) \wedge (z \vee x) = x.$

Take $y := y \vee x, \; u := y, \; v := z$ in 36, then use 24 with $x := y, \; y := x.$

39. $x \vee (((y \wedge x) \vee (z \wedge x)) \wedge u) = x.$

Take $v := x \vee (((y \wedge x) \vee (z \wedge x)) \wedge u)$ in 36, then use 25 with $x := x \vee (((y \wedge x) \vee (z \wedge x)) \wedge u), \; y := x.$

40. $((x \vee y) \wedge (y \vee z)) \wedge y = y.$

Identity 37 reduces to 40 by 33 with $x := y, \; y := x.$

41. $(((x \wedge y) \vee (z \wedge y)) \vee (((x \wedge y) \vee (z \wedge y)) \wedge u)) \wedge y = (x \wedge y) \vee (z \wedge y).$

Take $x := (x \wedge y) \vee (z \wedge y), \; y := u, \; z := (x \vee y) \wedge y$ in 38, then use 26 with $z := x, \; u := z.$

42. $(x \wedge y) \wedge x = x \wedge y$.

Take $x := x \wedge y$, $y := x$, $z := x$ in 33, then use 38 with $z := x \wedge y$.

43. $x \vee (((y \wedge x) \vee x) \wedge z) = x$.

Take $z := y \vee x$, $u := z$ in 39, then use 34 with $x := y$, $y := x$.

44. $x \vee ((y \wedge x) \wedge z) = x$.

Take $z := y$, $u := z$ in 39, then use 32 with $x := y \wedge x$.

45. $((x \wedge y) \vee y) \wedge y = (x \wedge y) \vee y$.

Take $x := (x \wedge y) \vee y$ in 30, then use 43 with $x := y$, $y := x$, $z := ((x \wedge y) \vee y) \vee z$.

46. $x \vee (y \wedge (z \wedge x)) = x$.

Take $y := z$, $z := y \wedge (z \wedge x)$ in 44, then use 35 with $x := z \wedge x$.

47. $(x \wedge y) \vee y = y$.

Identity 45 reduces to 47 by 34 with $x := x \wedge y$.

48. $x \vee (y \wedge (x \wedge z)) = x$.

Take $z := x \wedge z$ in 46, then use 42 with $y := z$.

49. $((x \wedge y) \vee (z \wedge y)) \vee y = y$.

Take $x := ((x \wedge y) \vee (z \wedge y)) \vee (((x \wedge y) \vee (z \wedge y)) \wedge u)$ in 47, then use 41.

50. $((x \wedge y) \vee (y \wedge z)) \vee y = y$.

Take $z := y \wedge z$ in 49, then use 42 with $x := y$, $y := z$.

Finally note that identities $28, 40, 48$ and 50 are $L_{24}^{\wedge}$, $L_{25}^{\wedge}$, $L_{24}^{\vee}$ and $L_{25}^{\vee}$, respectively.   □

McCune has now created Prover9 – a new software, an improved version of Otter, and he recommends using only Prover9.

So the (improper) variety **L** of all lattices can be characterized by a single axiom. Likewise, the trivial variety **T** of one-element lattices is characterized by the axiom $x = y$. These simple remarks cannot be improved:

Theorem 1.3.3. (McKenzie [1970]) *No non-trivial proper variety of lattices can be defined by a single identity.*

PROOF: Suppose that such a variety **K** is characterized by a single identity, which, by Lemma 3.1, is of the form $f(x, x_1, \ldots, x_n) = x$. Take a non-trivial lattice $L \in \mathbf{K}$. Then there exist $o, u \in L$ satisfying $o < u$ and the identity $f = x$ holds in the sublattice $\{o, u\}$ of L, therefore it is valid in every two-element lattice.

Now define $f_0(x, x_1, \ldots, x_n) = f(x, z, \ldots, z)$, where $z = x \wedge x_1 \wedge \ldots \wedge x_n$, and $f_1(x, x_1, \ldots, x_n) = f(x, u, \ldots, u)$, where $u = x \vee x_1 \vee \ldots \vee x_n$. Then both $f_0 = x$ and $f_1 = x$ are valid in any two-element lattice.

On the other hand $z < x < u$ because the variables $x, x_1, \ldots, x_n$ are distinct. Since lattice polynomials are isotone, the inequalities

$$(9) \qquad\qquad\qquad f_0 \leq f \leq f_1$$

hold in any lattice.

The validity of $f_0 = x$ in the lattice $\{z, x\}$ implies that $f_0 = z$ or $f_0 = x$, where in fact $x, x_1, \ldots, x_n$ are arbitrary. In other words we have $f_0 = z$ or $f_0 = x$ in every lattice. The former alternative applied to the lattice $\{z, x\}$ yields $x = z$, a contradiction. Therefore $f_0 = x$ in every lattice and similarly $f_1 = x$ in every lattice. This transforms (9) into $x \leq f \leq x$, that is, the identity $f = x$ holds in any lattice, which contradicts the initial assumption. $\qquad\qquad\qquad\qquad\qquad\qquad\qquad\qquad\qquad\qquad\qquad\qquad$ $\square$

Corollary 1.3.1. *A variety* **K** *of lattices can be defined by a single axiom iff either* **K**=**T** *(the trivial variety of one-element lattices), or* **K**=**L** *(the variety of all lattices).*

PROOF: This summarizes Theorem 3.3 and the comments preceding it.
$\qquad\qquad\qquad\qquad\qquad\qquad\qquad\qquad\qquad\qquad\qquad\qquad\qquad\qquad\qquad\qquad$ $\square$

In particular *the variety* **D** *of all distributive lattices is not one-based.* However we know by Corollary 2.3 that every finitely-based variety of lattices is two-based. We can apply the above idea of absorption law to prove that several varieties of enriched lattices obtained by adjoining 0 or 1 are one-based. This is because the property of an element being 0 or 1 can be captured by an absorption law: $x \vee 0 = x$ and $x \wedge 1 = x$, respectively. This idea can be streched further. An algebra $(L, \vee, \wedge, ')$ is a complemented lattice if and only if it is a lattice satisfying the absorption laws $x \vee (y \wedge y') = x$ and $x \wedge (y \vee y') = x$. Summarizing the previous results and anticipating results of later sections, we have the following: *the varieties of all lattices, trivial lattices, lattices with* least element 0, *lattices with* greatest element 1, *bounded lattices (with 0 and 1), ℓ-groups and Boolean algebras are one-based, while the varieties of semilattices, quasilattices and finitely-based lattices are two-based but not one-based.*

1.4. Defining Lattices by Other Tools

In this section we present definitions of lattices using other tools than the order $\leq$ or the operations $\vee, \wedge$, namely segments or lattice betweenness for lattices with least element 0, K-segments for arbitrary lattices, a quaternary operation for finitely definable equational classes of lattices, and a partially defined ternary operation for lattices wirh 0 and 1.

For every two elements a, b of a lattice L, the *segment* $[a, b]$ is defined by

(1) $$[a, b] = \{x \in L \mid a \wedge b \leq x \leq a \vee b\} \,.$$

If $a \leq b$, the segment $[a, b]$ reduces to the *interval*

(2) $$[a, b] = \{x \in L \mid a \leq x \leq b\} \,.$$

Theorem 1.4.1. (Nishigōri [1954])[†] α) *An algebra* $(L, \vee, \wedge, 0)$ *of type* (2,2,0) *is a lattice with least element* 0 *if and only if with each couple* $(x, y) \in L^2$ *is associated a nonempty subset* $[x, y]$ *of* L *such that the following conditions are fulfilled: for every* $x, y, u, v \in L$,

L$_{38}$ $[x, x] = \{x\}$,

L$_{39}$ $[0, x] \cap [0, y] = [0, x \wedge y]$,

L$_{40}$ $[0, x] \subseteq [0, x \vee y]$,

L$_{41}$ $[0, y] \subseteq [0, x \vee y]$,

L$_{42}$ $[0, x] \subseteq [0, v] \,\&\, [0, y] \subseteq [0, v] \implies [0, x \vee y] \subseteq [0, v]$,

L$_{43}$ $[x, y] \subseteq [u, v] \iff [0, u] \cap [0, v] \subseteq [0, x] \cap [0, y] \,\&\, [0, x \vee y]$
$\phantom{L_{43} \quad} \subseteq [0, u \vee v]$.

β) *When this holds, the "segments"* $[a, b]$ *coincide with* (1) *and the order relation of the lattice is given by*

(3) $$x \leq y \iff [0, x] \subseteq [0, y] \,.$$

PROOF: Given a lattice L with 0, definition (1) immediately implies (3) and L$_{38}$ − L$_{43}$.

Conversely, suppose L$_{38}$ − L$_{43}$ hold and define the relation $\leq$ by (3). Then $\leq$ is reflexive and transitive, and

(4) $$x \wedge y \leq x \,\&\, x \wedge y \leq y \,,$$

(5) $$x \leq x \vee y \,\&\, y \leq x \vee y \,,$$

(6) $$x \leq v \,\&\, y \leq v \implies x \vee y \leq v \,,$$

while L$_{43}$ can be written in the form

(7) $$[x, y] \subseteq [u, v] \iff u \wedge v \leq x \wedge y \,\&\, x \vee y \leq u \vee v \,.$$

[†]L$_{42}$ is a weakening of the original axiom.

Then

(8) $$v \leq x \; \& \; v \leq y \Longrightarrow v \leq x \wedge y \,,$$

because the hypothesis implies $[0, v] \subseteq [0, x] \cap [0, y] = [0, x \wedge y]$.
Taking $y := v := x$ in (4), (8), (5) and (6), we obtain

(9) $$x \wedge x \leq x \leq x \wedge x \; \& \; x \leq x \vee x \leq x \,.$$

It follows from L_{37} and L_{38} that

$$\{0\} \cap [0, x] = [0, 0] \cap [0, x] = [0, 0 \wedge x] \neq \varnothing \,,$$

hence $0 \in [0, x]$, therefore $[0, 0] = \{0\} \subseteq [0, x]$, that is,

(10) $$0 \leq x \,.$$

It remains to prove (1) and antisymmetry.
Taking in (7) $y := x$ and using L_{38} and (9), we infer

$$x \in [u, v] \Longleftrightarrow [x, x] \subseteq [u, v] \Longleftrightarrow u \wedge v \leq x \wedge x \; \& \; x \vee x \leq u \vee v$$

$$\Longrightarrow u \wedge v \leq x \; \& \; x \leq u \vee v \Longrightarrow u \wedge v \leq x \wedge x \; \& \; x \vee x \leq u \vee v \Longleftrightarrow x \in [u, v] \,,$$

therefore we get (1) in the form

$$x \in [u, v] \Longleftrightarrow u \wedge v \leq x \; \& \; x \leq u \vee v \,.$$

Finally it follows from (9) that

$$x \leq u \Longrightarrow x \vee x \leq x \leq u \leq u \vee u \,,$$

$$u \leq x \Longrightarrow u \wedge u \leq u \leq x \leq x \wedge x \,,$$

hence

$$x \leq u \; \& \; u \leq x \Longrightarrow x \vee x \leq u \vee u \; \& \; u \wedge u \leq x \wedge x \,.$$

On the other hand we infer from (7) with $y := x$ and $v := u$ that

$$x = u \Longleftrightarrow \{x\} \subseteq \{u\} \Longleftrightarrow u \wedge u \leq x \wedge x \; \& \; x \vee x \leq u \vee u \,,$$

therefore

(11) $$x \leq u \; \& \; u \leq z \Longrightarrow x = y \,.$$

$\square$

Smiley and Transue [1943] have worked with the concept of *lattice betweenness*, which is the ternary relation defined by

(12) $$axb \Longleftrightarrow (a \wedge x) \vee (x \wedge b) = x = (a \vee x) \wedge (x \vee b) \,.$$

This concept originates in the theory of metric lattices, initiated by Glivenko. In a lattice endowed with a distance function d, an element x is said to be between two elements a, b if $d(a, x) + d(x, b) = d(a, b)$. Glivenko proved that this happens if and only if the above property axb holds. But definition (12) makes sense in any lattice and the following easy consequences of (12) have been considered by Smiley and Transue:

L_{44} $abc \Longleftrightarrow cba$,

L_{45} $abc \ \& \ acb \Longleftrightarrow b = c$,

L_{46} $abc \ \& \ adb \Longrightarrow dbc$,

and if the lattice has least element 0, then

(13) $x \leq y \Longleftrightarrow 0xy$,

L_{47} $0bc \ \& \ 0dc \ \& \ bxd \Longrightarrow 0xc$,

L_{48} $(a \vee b)a(a \wedge b)$, $(a \vee b)b(a \wedge b)$, $a(a \vee b)b$, $a(a \wedge b)b$, $0(a \wedge b)(a \vee b)$,

L_{49} $x \wedge (p \vee c) = (p \wedge x) \vee (x \wedge c) \ \& \ x \vee (p \wedge c) = (p \vee x) \wedge (x \vee c) \Longrightarrow$

                    $(pbc \ \& \ pdc \ \& \ bxd \Longrightarrow pxc)$.

Property L_{44} is immediate. If $abc \ \& \ acb$ then $a \wedge b \leq (a \vee c) \wedge (c \vee b) = c$, hence $b = (a \wedge b) \vee (b \wedge c) \leq c$ and similarly $c \leq b$, hence $b = c$, thus proving L_{45}. If $abc \ \& \ adb$ then from $(a \vee d) \wedge (d \vee b) = d$ we infer $(a \vee d) \wedge b = b \wedge d$, then

$$b = (a \wedge b) \vee (b \wedge c) \leq ((a \vee d) \wedge b) \vee (b \wedge c) = (b \wedge d) \vee (b \wedge c) \leq b ,$$

hence $b = (d \wedge b) \vee (b \wedge c)$ and similarly $b = (d \vee b) \wedge (b \vee c)$. Therefore L_{46} holds.

One checks readily that $0xy \Longleftrightarrow x = x \wedge y$, which proves (13). Now L_{47} reads

$$b \leq c \ \& \ d \leq c \ \& \ bxd \Longrightarrow x \leq c$$

and in fact the hypotheses imply

$$x = (b \wedge x) \vee (x \wedge d) \leq b \vee d \leq c .$$

Checking L_{48} is routine. To prove L_{49} note that

$$pbc \Longrightarrow b \leq p \vee c , \quad pdc \Longrightarrow d \leq p \vee c , \quad bxd \Longrightarrow x \leq b \vee d ,$$

hence $x \leq p \vee c$, and $x \geq p \wedge c$ by duality. Therefore $x = x \wedge (p \vee c) = (p \wedge x) \vee (x \wedge c)$ and $x = x \vee (p \wedge c) = (p \vee x) \wedge (x \vee c)$.

Smiley and Transue (op. cit.) have proved a theorem which is slightly more general than the following

Theorem 1.4.2. α) *An algebra* $(L, \vee, \wedge, 0)$ *of type* $(2,2,0)$ *is a lattice with first element* 0 *if and only if it is endowed with a ternary relation satisfying* $L_{44} - L_{49}$.

β) *When this holds, the "lattice betweenness" coincides with* (12), *while the order relation of the lattice is given by* (13).

See also Blumenthal and Bumcrot [1962].

Starting from lattice betweenness, Kolibiar [1958] introduced a concept which we will call K-segment. Given a lattice L and two elements $a, b \in L$, let

$$(14) \qquad\qquad B(a,b) = \{x \in L \mid axb\}$$

be the set of elements that are *lattice-between* a and b. Note that

$$(15) \qquad\qquad a, b, a \wedge b, a \vee b \in B(a,b),$$

$$(16) \qquad\qquad B(a,b) \subseteq [a,b] \subseteq B(a \wedge b, a \vee b).$$

Let us refer to a subset $F \subseteq L$ satisfying $a, b \in F \implies B(a,b) \in F$ as a *lattice-convex* set. Let $X \mapsto \overline{X}$ denote the closure operator associated with the Moore family of lattice-convex sets. Then

$$(17) \qquad\qquad \overline{B(a,b)} = [a,b].$$

For if $x, y \in [a,b]$ then $a \wedge b \leq x \wedge y \leq x \vee y \leq a \vee b$, which, using (16), implies $B(x,y) \subseteq [x,y] \subseteq [a,b]$. Therefore $[a,b]$ is a lattice-convex set which includes $B(a,b)$. Suppose $B(a,b) \subseteq F$, where F is a lattice-convex set. If $x \in [a,b]$ then $x \in B(a \wedge b, a \vee b)$ by (16); but $a \wedge b, a \vee b \in F$ by (15), therefore $x \in F$. This completes the proof of (17).

If the set $B(a,b)$ is closed, then we put

$$(18) \qquad\qquad K(a,b) = B(a,b) = [a,b]$$

and refer to this set as the *K-segment* determined by (a,b); otherwise the couple $K(a,b)$ does not determine a K-segment.

Kolibiar (op. cit.) has considered the following properties of segments, lattice betweenness and K-segments:

L_{50} $[a,b] \cap [b,c] \cap [c,a] \neq \varnothing$,

L_{51} $axb \Longleftrightarrow [a,x] \cap [b,x] = \{x\}$,

L_{52} every three elements a, b, c are contained in a K-segment;

L_{53} with each K-segment $K(a,b)$ is associated an "oriented" K-segment $K(o,u)$, such that

       (i) $K(a,b) = K(o,u)$, and

       (ii) for every two oriented K-segments $K(o,u), K(o',u')$, if $K(o,u) \subseteq K(o',u')$ and the K-oriented segment $K(o',u)$ exists, then $o \in K(o',u)$.

Property L_{50} follows from

$$[a,b] \cap [b,c] \cap [c,a] = [(a \wedge b) \vee (b \wedge c) \vee (c \wedge a), (a \vee b) \wedge (b \vee c) \wedge (c \vee a)] ,$$

and this also proves L_{52}, because every segment is closed by (17).

Further, note that since

(19) $(a \wedge x) \vee (b \wedge x) \leq x \leq (a \vee x) \wedge (b \vee x)$,

it follows by comparison with (12) that in fact

(20) $axb \Longleftrightarrow (a \wedge x) \vee (b \wedge x) = (a \vee x) \wedge (b \vee x)$.

On the other hand,

(21) $[a,x] \cap [b,x] = [(a \wedge x) \vee (b \wedge x), (a \vee x) \wedge (b \vee x)]$,

hence (20) says that axb holds if and only if the segment (21) is a singleton, and in view of (19) this singleton is $\{x\}$, thus proving L_{51}.

To prove L_{53} note that if $o \leq u$ then

$$oxu \Longleftrightarrow (o \vee x) \wedge (x \vee u) = x = (o \wedge x) \vee (x \wedge u)$$

$$\Longleftrightarrow o \vee x = x = x \wedge u \Longleftrightarrow o \leq x \leq u .$$

In other words, $o \leq u \Longrightarrow B(o,u) = [o,u]$ and we can take as oriented K-segments, the segments $[o,u]$ with $o \leq u$. So, given $K(a,b)$, we have $K(a,b) = K(o,u)$ with

(22) $o = a \wedge b$, $u = a \vee b$.

Besides, a stronger form of L_{53}(ii) holds: if $K(o,u)$ and $K(o',u')$ are oriented K-segments satisfying $K(o,u) \subseteq K(o',u')$, then $o' \leq o \leq u \leq u'$, showing that the oriented K-segment $K(o',u)$ exists and $o \in K(o',u)$.

Theorem 1.4.3. (Kolibiar [1958]) *Let L be a set endowed with a function $[\ ] : L^2 \longrightarrow \mathcal{P}(L)$, a partial function $K : L^2 \overset{\circ}{\longrightarrow} \mathcal{P}(L)$ and a ternary relation $\beta \subseteq L^3$; let axb stand for $(a, x, b) \in \beta$. Then L is a lattice $(L, \vee, \wedge, \leq)$ in which relations* (1), (12), (14), (18) *and* (22) *hold if and only if it satisfies* $L_{50} - L_{53}$.

We omit the more elaborated rest of the proof. See also Hedlíková and Katriňák [1991] and Ploščica [1996].

Theorem 1.4.4. *Let $\mathbf{K}$ be the class of lattices defined within the class of all lattices by an identity $f = g$, where f and g are lattice polynomials. Let $(L, \vee, \wedge)$ be an algebra of type* (2,2) *and define*

(23) $$q(a, b, c, d) = ((a \vee b) \wedge c) \vee (a \wedge d) .$$

Further let Q be the equation obtained from $L_{28}^{\vee}$ *by the transformations*

(24) $$a \vee b = q(b, a, a, b) ,$$

(25) $$a \wedge b = q(b, b, a, a) .$$

Then $(L, \vee, \wedge) \in \mathbf{K}$ if and only if (L, q) satisfies Q and

L_{54} $\qquad q(a, x, q(a, a, z, t), a) = a .$

PROOF: In view of Theorem 2.4, we have to prove that $(L, \vee, \wedge)$ satisfies $L_{27}^{\vee}$ and $L_{28}^{\vee}$ if and only if (L, q) satisfies L_{54} and Q.

Suppose $(L, \vee, \wedge)$ satisfies $L_{27}^{\vee}$ and $L_{28}^{\vee}$. Then $\vee$ and $\wedge$ coincide with the operations defined by (24) and (25) because

$$q(b, a, a, b) = ((b \vee a) \wedge a) \vee (b \wedge b) = a \vee b ,$$

$$q(b, b, a, a) = ((b \vee b) \wedge a) \vee (b \wedge a) = a \wedge b .$$

Therefore Q follows from $L_{28}^{\vee}$, while L_{54} is fulfilled because

$$q(a, x, q(a, a, z, t), a) = ((a \vee x) \wedge (q(a, a, z, t))) \vee (a \wedge a)$$

$$= ((a \vee x) \wedge (((a \vee a) \wedge z) \vee (a \wedge t))) \vee a = ((a \vee x) \wedge ((a \wedge z) \vee (a \wedge t))) \vee a$$

$$= ((a \wedge z) \vee (a \wedge t)) \vee a = a .$$

Conversely, suppose (L, q) satisfies Q and L_{54}. Then it follows from (24) and (25) that $(L, \vee, \wedge)$ satisfies $L_{28}^{\vee}$ and also $L_{27}^{\vee}$, because

$$(x \wedge y) \vee y = q(y, x \wedge y, x \wedge y, y) = q(y, x \wedge y, q(y, y, x, x), y) = y$$

by L_{54} with $a := y$, $x := x \wedge y$, $z := t := x$.   □

Corollary 1.4.1. $(L, \vee, \wedge)$ *is a lattice if and only if* (L, q) *satisfies* L_{54} *and the identity* Q' *obtained from* $L_{29}^{\vee}$ *by the transformation* (24), (25).

PROOF: From Theorem 4.4 and Corollary 2.2.   □

Corollary 1.4.2. *Every finitely definable class of lattices can be characterized in terms of a quaternary operation.*

PROOF: By Theorems 2.3 and 4.4.   □

It was shown by D. Kelly and Padmanabhan [1989] that lattices cannot be defined in terms of a ternary operation alone, because a ternary polynomial cannot express both the join and the meet.

However in a *bounded lattice*, that is, a lattice with 0 and 1, the ternary operation

$$(26) \qquad\qquad t(a, b, c) = (a \vee c) \wedge (b \vee (a \wedge c))$$

yields the lattice operations via formulae

$$(27) \qquad\qquad a \wedge c = t(a, 0, c), \quad a \vee c = t(a, 1, c).$$

Martin [1965] characterized lattices by the axioms

L_{55} $t(a, b, c) = t(c, b, a)$,

L_{56} $t(0, 1, a) = a$,

L_{57} $t(1, 0, a) = a$,

L_{58} $t(a, 1, t(b, 1, c)) = t(c, 1, t(b, 1, a))$,

L_{59} $t(a, 0, t(b, 0, c)) = t(c, 0, t(b, 0, a))$,

L_{60} $t(a, 0, t(a, 1, b)) = a$,

L_{61} $t(a, 1, t(a, 0, b)) = a$,

to the effect that in every lattice the operations (26) satisfy properties $L_{55} - L_{61}$ and conversely, every algebra (L, t) of type (3) which fulfils these axioms becomes a bounded lattice $(L, \vee, \wedge, 0, 1)$ with respect to the operations (26). See also the operation $s(x, y, z) = t(y, x, z)$ in Chapter 2, §2, and the median operation in Chapter 3, §3.

Kolibiar [1956a] characterized bounded lattices in terms of the partially defined ternary operation

$$(28) \qquad < a, b, c > = (a \wedge b) \vee (b \wedge c) \vee (c \wedge a) = (a \vee b) \wedge (b \vee c) \wedge (c \vee a),$$

where $< a, b, c >$ is not defined if the second equality in (28) does not hold, and which satisfies certain axioms. Conversely, his axioms imply only a weaker form of (28), namely

$$(a \wedge b) \vee (b \wedge c) \vee (c \wedge a) \leq \; < a, b, c > \; \leq (a \vee b) \wedge (b \vee c) \wedge (c \vee a) \; .$$

Katriňák [1961] has shown that two operations $<>$ and $<>'$ satisfying the axioms need not coincide even if they have the same domain of definition.

Other problems concern *self-dual varieties* of lattices. A variety of lattices is called self-dual provided the principle of duality holds in all of its members. For instance, the varieties of all lattices, modular lattices, distributive lattices and Boolean algebras (cf. next chapters) are self-dual. Clearly every self-dual variety can be defined by a self-dual system of axioms. It is natural to ask whether a given finitely-based self-dual variety of lattices can be defined by a single self-dual axiom relative to the class **L** of all lattices. D. Kelly and Padmanabhan [2002] proved that there are infinitely many varieties for which the answer is "yes" and infinitely many varieties which don't have such a single-axiom characterization. Let us say that these varieties are of the *first kind* and of the *second kind*, respectively.

On the other hand, a self-dual system of axioms need not be independent and an independent system need not be self-dual. Therefore it is natural to look for irredundant (i.e., independent) self-dual bases for finitely based self-dual lattice varieties. D. Kelly and Padmanabhan [2004] proved the following result. Associate with every finitely based self-dual lattice variety **V** a number n_0 defined as follows: $n_0 = 1$ for **T**, $n_0 = 2$ for **L**, $n_0 = 3$ if **V** is of the first kind, and $n_0 = 4$ if **V** is of the second kind. Then for every finitely based self-dual variety **V** and every $n \geq n_0$ there is an irredundant self-dual basis with n identities defining **V**.

Chapter 2

Modular Lattices

Modular lattices were introduced by Dedekind [1897] as an abstract characterization of the lattice of normal subgroups of a group; see Birkhoff [1948] for historic references. Modular lattices have also a strong connection with projective planes. Their systematic study was begun by Ore[1935].

In this chapter, after a short introduction of modular lattices as a subclass of the class of all lattices, we survey direct definitions of modular lattices, either in terms of disjunction and conjunction, or by other tools. The last section is devoted to the construction of self-dual equational bases for varieties of modular lattices. The final result is that any finitely-based self-dual variety of modular lattices can be defined as an equational class of lattices satisfying a single identity. This is a special case of a most general result on the so-called p-modular lattices, proved by D. Kelly and Padmanabhan [2002].

2.1. Modular Lattices within Lattices

A lattice $(L, \vee, \wedge)$ is said to be *modular* provided

$$(1) \qquad x \leq z \Longrightarrow x \vee (y \wedge z) = (x \vee y) \wedge z \ .$$

Note that the dual of condition (1), that is,

$$(1') \qquad x \geq z \Longrightarrow x \wedge (y \vee z) = (x \wedge y) \vee z \ ,$$

coincides with (1) due to commutativity. In other words, axiom (1) is self-dual, therefore the *principle of duality holds for modular lattices*.

Condition (1) is not an identity, but it is easy to find identities equivalent to (1) within the class of lattices, thus showing that *the class of modular lattices is a subvariety* of the class of all lattices. The most natural such identity is perhaps

$$(2) \qquad x \vee (y \wedge (x \vee z)) = (x \vee y) \wedge (x \vee z) \ .$$

For (1) implies (2) by taking $z := x \vee z$, while (2) implies

$$x \leq z \implies x \vee (y \wedge z) = x \vee (y \wedge (x \vee z)) = (x \vee y) \wedge (x \vee z) = (x \vee y) \wedge z \ .$$

In view of the principle of duality, axiom (2) is also equivalent to its dual

(2') $$x \wedge (y \vee (x \wedge z)) = (x \wedge y) \vee (x \wedge z) \ .$$

Since the class of modular lattices is self-dual, it is natural to ask whether it can also be defined by a single self-dual identity. Such an identity is well known. It is sometimes called the *shearing identity*:

(3) $$(x \wedge (y \vee z)) \vee (y \wedge z) = (x \vee (y \wedge z)) \wedge (y \vee z) \ .$$

It is because of the name "shearing" that in subsequent pages we use the notation $s(x, y, z)$ (first occurrence in identity (16)). Clearly (1) implies (3) by taking $x := y \wedge z$, $y := x$, $z := y \vee z$, while (3) implies

$$y \leq z \implies y \vee (x \wedge z) = (y \wedge z) \vee (x \wedge (y \vee z)) = (x \vee (y \wedge z)) \wedge (y \vee z) = (y \vee x) \wedge z \ .$$

Therefore, by adding the self-dual axiom (3) to any self-dual system of axioms defining lattices one obtains a *self-dual system of axioms for modular lattices*. See also the end of §2, where identity (3) appears in the form (2.17).

We recall that a *sublattice* of a lattice $(L, \vee, \wedge)$ is a subset $S \subseteq L$ such that for any $x, y \in S$ it follows that $x \vee y$, $x \wedge y \in S$. The following well-known characterization of modular lattices is very useful:

Theorem 2.1.1. *A lattice is modular if and only if it does not include a sublattice of the form in Fig.1.*

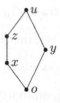

Fig.1

Comment Fig.1 is a *Hasse diagram*, in which a segment with ends a and b, a below b, indicates that b *covers* a, that is, the following two conditions are fulfilled: $a < b$ (meaning $a \leq b$ and $a \neq b$) and there is no element x such that $a < x < b$. The lattice in Fig.1 is known as N5.

PROOF: A lattice which includes a sublattice of the form in Fig.1 is not modular, because $x < z$ but $x \vee (y \wedge z) = x \vee o = x$, while $(x \vee y) \wedge z = u \wedge z = z$.

Conversely, suppose there exist x, y, z such that $x \leq z$ but $x \vee (y \wedge z) \neq (x \vee y) \wedge z$. Since $x \vee (y \wedge x) = x = (x \vee y) \wedge x$, it follows that $x \neq z$, hence $x < z$. Since $x \vee (y \wedge z) \leq (x \vee y) \wedge z$, it follows that $x \vee (y \wedge z) < (x \vee y) \wedge z$. Hence $x \leq y$ would imply $y \wedge z < y \wedge z$, a contradiction, while $y \leq x$ would imply the contradiction $x < x \wedge z$. Therefore x and y are incomparable. Setting $x \wedge y = o$, it follows that $o < x$ and $o < y$. Similarly, setting $z \vee y = u$, it follows that $z < u$ and $y < u$. Thus we have obtained the sublattice in Fig.1. □

Corollary 2.1.1. (Kurosh [1935], Ore [1935]) *A lattice is modular if and only if*

(4) $$x \leq y \ \& \ \exists z \, (x \wedge z = y \wedge z \, \& \, x \vee z = y \vee z) \Longrightarrow x = y \,.$$

PROOF: If the lattice is modular, the left-hand side of (4) implies

$$x = x \vee (x \wedge z) = x \vee (z \wedge y) = (x \vee z) \wedge y = (y \vee z) \wedge y = y \,.$$

If the lattice is not modular, the sublattice in Fig.1 shows that condition (4) fails for $y := z$ and $z := y$. □

Corollary 2.1.2. *A lattice is modular if and only if it satisfies the identity*

(5) $$x \vee (y \wedge (x \vee z)) = x \vee (z \wedge (x \vee y)) \,.$$

PROOF: If the lattice is modular, then from $x \leq x \vee z$ and $x \leq x \vee y$ we infer

$$x \vee (y \wedge (x \vee z)) = (x \vee y) \wedge (x \vee z) = (x \vee z) \wedge (x \vee y) = x \vee (z \wedge (x \vee y)) \,,$$

while if the lattice is not modular then (5) fails for the sublattice in Fig.1:

$$x \vee (y \wedge (x \vee z)) = x \vee (y \wedge z) = x \vee o = x \ \text{and} \ x \vee (z \wedge (x \vee y)) = x \vee z = z \,.$$

□

Remark 2.1.1. Nishigōri [1954] has translated the modularity condition in terms of segments.

Theorem 2.1.2. *There is no system of several mutually independent identities characterizing modular lattices within the class of all lattices.*

PROOF: Suppose Σ would be such a system. Then some member of Σ fails in the lattice in Fig.1, say $f = g$. Hence $f = g$ already characterizes

modular lattices, therefore the existence of other members of Σ contradicts
its independence.   □

2.2. Defining Modular Lattices in Terms of the Operations ∨ and ∧ or by Other Tools

The first direct definition of modular lattices, that is, not including a system
of axioms for lattices, is due to Kolibiar [1956b], namely the independent
system

$$\mathbf{M}_1 = \{\mathrm{M}_1, \mathrm{M}_2\} \, ,$$

where

M$_1$ $((x \wedge y) \wedge z) \vee (x \wedge t) = ((t \wedge x) \vee (z \wedge y)) \wedge x$,

M$_2$ $(x \vee (y \wedge y)) \wedge y = y$.

Then Riečan [1957] devised the system

$$\mathbf{M}_2 = \{\mathrm{M}_3, \mathrm{M}_4, \mathrm{M}_5\} \, ,$$

where

M$_3$ $(x \vee y) \wedge y = y$,

M$_4$ $(x \vee y) \vee z = x \vee (y \vee z)$,

M$_5$ $(x \wedge y) \vee (x \wedge z) = ((z \wedge x) \vee y) \wedge x$,

and the independent system

$$\mathbf{M}_3 = \{\mathrm{M}_5, \mathrm{M}_6\} \, ,$$

where

M$_6$ $(x \vee (y \vee z)) \wedge z = z$.

Sobociński [1975a], being not aware of systems $\mathbf{M}_2, \mathbf{M}_3$, obtained the independent system

$$\mathbf{M}_4 = \{\mathrm{M}_7, \mathrm{M}_8\} \, ,$$

where

M$_7$ $(x \wedge y) \vee (x \wedge z) = ((z \wedge x) \vee (y \vee y)) \wedge x$,

M$_8$ $(z \vee (y \vee x)) \wedge x = x$,

then he discovered Riečan's paper and obtained another system $\mathbf{M}_2'$ by
replacing M$_4$ by its dual in $\mathbf{M}_2$; cf. Sobociński [1976a]. He also conjectured
that in $\mathbf{M}_3$ axiom M$_6$ cannot be replaced by $(y \vee z) \wedge z = z$, which was
confirmed by Sudkamp [1976]. Quite independently, Ponticopoulos [1964]
suggested a 3-axiom system with 12 variables.

Theorem 2.2.1. $\mathbf{M_3}$ *is an independent system which defines modular lattices.*

PROOF: In every lattice condition $\mathbf{M_6}$ is fulfilled, while $\mathbf{M_5}$ is equivalent to the modularity condition $(2')$. Conversely, suppose an algebra $(L, \vee, \wedge)$ satisfies $\mathbf{M_3}$. In view of the above remark, if we prove that L is a lattice, it will follow that it is a modular one.

By applying $\mathbf{M_5}$ with $y := y \vee x$, then $\mathbf{M_6}$ with $x := z \wedge x$, $z := x$, we obtain

$$(1) \qquad (x \wedge (y \vee x)) \vee (x \wedge z) = ((z \wedge x) \vee (y \vee x)) \wedge x = x \; .$$

By applying (1), then $\mathbf{M_6}$ with $x := y$, $y := x \wedge (y \vee x)$, $z := x \wedge z$, we infer

$$(2) \qquad (y \vee x) \wedge (x \wedge z) = (y \vee ((x \wedge (y \vee x)) \vee (x \wedge z))) \wedge (x \wedge z) = x \wedge z \; .$$

By using in turn (1), then $\mathbf{M_5}$ with $x := y \vee x$, $y := x \wedge z$, $z := x$, and (2), we get

$$x \wedge (y \vee x) = ((x \wedge (y \vee x)) \vee (x \wedge z)) \wedge (y \vee x)$$

$$= ((y \vee x) \wedge (x \wedge z)) \vee ((y \vee x) \wedge x) = (x \wedge z) \vee ((y \vee x) \wedge x) \; ;$$

we have thus proved that

$$(3) \qquad (x \wedge z) \vee ((y \vee x) \wedge x) = x \wedge (y \vee x) \; .$$

Further we apply (1) and (3), both with $y := x$, $z := x \vee x$, which yields

$$x = (x \wedge (x \vee x)) \vee (x \wedge (x \vee x)) = (x \wedge (x \vee x)) \vee ((x \wedge (x \vee x)) \vee ((x \vee x) \wedge x)) \; ,$$

that is, $x = y \vee (y \vee z)$, where we have set $y = x \wedge (x \vee x)$ and $z = (x \vee x) \wedge x$. By applying $\mathbf{M_6}$ we get $x \wedge z = (y \vee (y \vee z)) \wedge z = z$, that is,

$$(4) \qquad x \wedge ((x \vee x) \wedge x) = (x \vee x) \wedge x \; .$$

The next computation uses in turn (3), (4), $\mathbf{M_5}$ with $y := x \vee x$, $z := (x \vee x) \wedge x$ and $\mathbf{M_6}$ with $x := ((x \vee x) \wedge x) \wedge x$, $y := z := x$:

$$x \wedge (x \vee x) = (x \wedge (x \vee x)) \vee ((x \vee x) \wedge x) = (x \wedge (x \vee x)) \vee (x \wedge ((x \vee x) \wedge x))$$

$$= ((((x \vee x) \wedge x) \wedge x) \vee (x \vee x)) \wedge x = x \; ;$$

we have thus proved that

$$(5) \qquad x \wedge (x \vee x) = x \; .$$

A new application of (1) yields

$$(x \wedge (x \vee x)) \vee (x \wedge (x \vee x)) = x$$

and taking into account (5) we obtain

(6) $$x \vee x = x .$$

It follows from M_6 and (6) that

(7) $$(x \vee y) \wedge y = y ,$$

then (6) and (7) yield

(8) $$y \wedge y = y .$$

Properties (8), M_5 and (7) imply

(9) $$x \vee (x \wedge z) = (x \wedge x) \vee (x \wedge z) = ((z \wedge x) \vee x) \wedge x = x .$$

From (7) and (9) we infer

(10) $$(x \vee y) \vee y = (x \vee y) \vee ((x \vee y) \wedge y) = x \vee y .$$

Further we apply in turn (8), (10), (8), M_5 with $x := z := x \vee y$, then (7) and (8):

$$x \vee y = (x \vee y) \wedge (x \vee y) = ((x \vee y) \vee y) \wedge (x \vee y)$$

$$= (((x \vee y) \wedge (x \vee y)) \vee y) \wedge (x \vee y) = ((x \vee y) \wedge y) \vee ((x \vee y) \wedge (x \vee y)) = y \vee (x \vee y) ,$$

hence

(11) $$y \vee (x \vee y) = x \vee y .$$

From (9), (11) and (9) we obtain

(12) $$(x \wedge y) \vee x = (x \wedge y) \vee (x \vee (x \wedge y)) = x \vee (x \wedge y) = x ,$$

while from (6), M_5 and (12) we get

$$x \wedge y = (x \wedge y) \vee (x \wedge y) = ((y \wedge x) \vee y) \wedge x = y \wedge x ,$$

that is,

(13) $$x \wedge y = y \wedge x .$$

It follows from (8), M_5, (8) and (12) that

(14) $$(x \vee y) \wedge x = ((x \wedge x) \vee y) \wedge x = (x \wedge y) \vee (x \wedge x) = (x \wedge y) \vee x = x .$$

Now we apply (14), (7), M_5, (13) and (7):

$$x \vee y = ((x \vee y) \wedge x) \vee ((x \vee y) \wedge y) = ((y \wedge (x \vee y)) \vee x) \wedge (x \vee y)$$

$$= (((x \vee y) \wedge y) \vee x) \wedge (x \vee y) = (y \vee x) \wedge (x \vee y) \,,$$

hence

$$x \vee y = (y \vee x) \wedge (x \vee y) \,,$$

therefore

$$y \vee x = (x \vee y) \wedge (y \vee x) \,,$$

and taking into account (13) it follows that

(15) $$x \vee y = y \vee x \,.$$

Summarizing, we have proved that M_5 and M_6 imply the commutativity of the two operations, i.e., (15), (13), their idempotency (6), (8) and the two absorption laws (12), (7). We are going to use freely these properties in proving the associativity of the two operations.

Set

$$x \vee (y \vee z) = a$$

and note that $a \wedge x = ((y \vee z) \vee x) \wedge x = x$ by (7) and $a \wedge y = y$ by M_6, hence M_5 implies

$$x \vee y = (a \wedge x) \vee (a \wedge y) = a \wedge (x \vee (y \wedge a)) = a \wedge (x \vee y) \,,$$

therefore, taking into account that $a \wedge z = z$ by M_6 and using M_5, we have

$$(x \vee y) \vee z = (a \wedge (x \vee y)) \vee (a \wedge z)$$

$$= a \wedge ((x \vee y) \vee (z \wedge a)) = a \wedge ((x \vee y) \vee z) \,.$$

Thus

$$(x \vee y) \vee z = (x \vee (y \vee z)) \wedge ((x \vee y) \vee z) \,,$$

which implies

$$x \vee (y \vee z) = (z \vee y) \vee x = (z \vee (y \vee x)) \wedge ((z \vee y) \vee x)$$

$$= (x \vee (y \vee z)) \wedge ((x \vee y) \vee z) = (x \vee y) \vee z \,.$$

We have thus proved that M_5 and M_6 imply the associativity M_4. Therefore, in order to prove the dual of M_4 it suffices to show that the duals of M_5 and M_6 are also valid. But

$$(x \vee y) \wedge (x \vee z) = ((x \wedge (x \vee z)) \vee y) \wedge (x \vee z)$$

$$= ((x \vee z) \wedge y) \vee ((x \vee z) \wedge x) = x \vee (y \wedge (z \vee x))$$

by M_5 with $x := x \vee z$, $z := x$, while M_4 implies

$$(x \wedge (y \wedge z)) \vee z = (x \wedge (y \wedge z)) \vee ((y \wedge z) \vee z)$$

$$= ((x \wedge (y \wedge z)) \vee (y \wedge z)) \vee z = (y \wedge z) \vee z = z \,.$$

Finally consider a set $\{a, b\}$. The operations $x \vee y = x \wedge y = a$ fulfil M_5 but not M_6, while the operations $x \vee y = x \wedge y = y$ satisfy M_6 but not M_5.
□

A somewhat surprising independent system was given by Vaida [1957], namely

$$\mathbf{M}_5 = \{M_9, M_{10}\} \,,$$

where

$M_9 \qquad x \wedge ((x \vee y) \vee z) = x$

and M_{10} is an ingenious translation of (1):

$M_{10} \qquad less(x, z) \implies z \wedge (y \vee x) = x \vee (y \wedge z) \,,$

where we have used the notation

$less(x, y) \iff x = z$ or $\exists z'\ x = z' \wedge z$ or $\exists z'\ x = z \wedge z'$ or $\exists z'\ z = z' \vee x \,.$

Kolibiar [1956b] characterized the class of modular lattices with 1 by the system

$$\mathbf{M}_6 = \{M_{11}, M_{12}, M_{13}\} \,,$$

where

$M_{11} \qquad ((x \wedge y) \wedge z) \vee (x \wedge t) = ((t \wedge x) \vee (z \wedge y)) \wedge x \,,$

$M_{12} \qquad x \vee 1 = 1 \,,$

$M_{13} \qquad 1 \wedge x = x \,,$

while Tamura [1975] improved this result by showing that the system

$$\mathbf{M}_7 = \{M_{14}, M_{15}\} \,,$$

where

M_{14} $(x \lor (y \land y)) \land y = y$,

M_{15} $(((0 \lor x) \land y) \land z) \lor (x \land t) = ((t \land x) \lor (z \land y)) \land x$,

defines modular lattices with 0. Sobociński [1979] defined modular lattices with 0 and 1 by a system with two identities again, having 8 and 3 variables, respectively.

The relation of lattice betweenness

$$(1.4.12) \qquad axb \iff (a \land x) \lor (x \land b) = x = (a \lor x) \land (x \lor b) ,$$

already introduced in Chapter 1, was also used for defining modular lattices. Smiley and Transue [1943] proved the following

Theorem 2.2.2. *Let* $(L, \lor, \land, 0)$ *be an algebra of type* $(2,2,0)$ *endowed with a ternary relation* $(4.1.12)$ *satisfying* L_{44}, L_{45}, L_{46} *and*

M_{16} $abc \,\&\, adb \implies adc$.

Then: $\alpha)$ L *is a modular lattice with least element 0 and order relation*

$$a \leq b \iff 0ab \text{ or } a = b$$

if and only if it satisfies L_{48} *and* L_{47} *or* L_{48} *and*

M_{17} $0bc \,\&\, 0bd \,\&\, cxd \implies 0bx$.

$\beta)$ *When the foregoing holds,* axb *is the lattice betweenness* $(1.4.12)$.

L.M. Kelly [1952] extended the above theorem to arbitrary modular lattices, while Kolibiar [1958] characterized modular lattices in terms of K-segments.

Setting

$$(16) \qquad s(x, y, z) = ((y \land z) \lor x) \land (y \lor z) ,$$

the characterization (1.3) of modularity within **L** can be written in the form

$$(17) \qquad s(x, y, z) = \tilde{s}(x, y, z) ,$$

where $\tilde{\ }$ is the operation of taking the dual. Several variations on this theme have been elaborated in the literature, as we are going to show in the remainder of this chapter.

Theorem 2.2.3. (Hashimoto [1951]) *In a bounded modular lattice* $(L, \lor, \land, 0, 1)$ *the ternary operation* (16) *satisfies* (17) *and*

M_{18} $\qquad$ $s(0, x, 1) = x$,

M_{19} $\qquad$ $s(x, y, y) = y$,

M_{20} $\qquad$ $s(s(x, v, w), y, s(z, v, w)) = s(x, s(y, w, v), s(z, w, v))$,

(18) $\qquad\qquad$ $x \vee y = s(x, 1, y)$, $x \wedge y = s(x, 0, y)$.

Conversely, if an algebra $(L, s, 0, 1)$ *of type* (3,0,0) *satisfies* $M_{18} - M_{20}$, *then* $(L, \vee, \wedge, 0, 1)$, *where* $\vee$ *and* $\wedge$ *are defined by* (18), *is a bounded modular lattice in which* (16) *and* (17) *hold.*

Kolibiar and Marcisová [1974] proved a similar theorem, in which axiom M_{20} is replaced by the axioms

M_{21} $\qquad$ $s(x, y, z) = s(x, z, y)$,

M_{22} $\qquad$ $s(s(x, v, z), y, z) = s(x, s(y, v, z), z)$.

We have seen in Chapter 1, §4, the characterization of lattices given by Martin [1965] in terms of the operation $t(x, y, z) = (x \vee z) \wedge (y \vee (x \wedge z))$; note that $t(x, y, z) = s(y, x, z)$ due to the commutativity of $\vee$ and $\wedge$. In the same paper, Martin characterized modular lattices by axioms L_{55}, L_{58}, L_{59} and

M_{23} $\qquad$ $t(x, z, x) = x$,

M_{24} $\qquad$ $t(t(x, 0, z), 1, t(y, 0, t(x, 1, z))) = t(x, y, z)$,

M_{25} $\qquad$ $t(t(x, 1, z), 0, t(y, 1, t(x, 0, z))) = t(x, y, z)$.

It follows from Theorem 3 in D. Kelly and Padmanabhan [1989] that a ternary lattice term p yields the join if and only if $p \geq t \vee (y \wedge z)$ for a permutation (t, y, z) of its variables. Therefore no ternary term p can yield both the join and the meet, because the above inequality together either with $p \leq t \wedge (y \vee z)$ or with $p \leq y \wedge (t \vee z)$, yield $t \leq y \vee z$, which need not hold.

So, while a 4-ary lattice term, e.g. $(x \wedge y) \vee (z \wedge u)$, can yield both $\vee$ and $\wedge$, a ternary lattice term can do this only with the aid of the constants 0 and 1.

The following result is also relevant:

Theorem 2.2.4. (Padmanabhan and Penner [2004]) *Let L be a subdirectly irreducible lattice. If $p(x, y, z)$ is an essentially ternary term such that $p(x, c, y)$ is a semilattice operation on L, then either L is bounded above with $c = 1$ and $p(x, 1, y) = x \vee y$, or L is bounded below with $c = 0$ and $p(x, 0, y) = x \wedge y$.*

2.3. Self-Dual Equational Bases for Modular Varieties

In this section we consider a non-trivial finitely-based equational class **K** of modular lattices. In view of Corollary 1.2.3, **K** can be characterized by two identities and according to Theorem 1.3.3, **K** is not one-based. However the number $n + 5$ of variables occurring in axiom L_{28} of Corollary 1.2.3 can be improved: we provide below a 3-basis involving $n + 3$ variables. Then we explore the possibility of obtaining self-dual equational bases of **K**; in particular we obtain a single-identity characterization relative to *the class* **M** *of modular lattices*.

To state our first result we use Theorem 1.2.3 and suppose, without loss of generality, that **K** is defined by a single identity $f = g$ along with the axioms of modular lattices. Let us introduce the following axiom:

(1) $$((x \wedge f) \wedge z) \vee (x \wedge t) = ((t \wedge x) \vee (z \wedge g)) \wedge x ,$$

where the variables x, z, t do not occur in f or g.

Theorem 2.3.1. *The following system defines* **K**: (1) *and*

$L_7^\vee$ $(x \wedge y) \vee y = y$,

$L_7^\wedge$ $(x \vee y) \wedge y = y$.

PROOF: If $L \in$ **K** then $L_7^\vee$ and $L_7^\wedge$ are fulfilled and taking $y := f = g$ in M_1 we obtain (1). Conversely, suppose L satisfies (1), $L_7^\vee$ and $L_7^\wedge$. According to a well-known argument (cf. L_1 in Ch.1,§2), the two absorption laws $L_7^\vee$ and $L_7^\wedge$ imply idempotency: $y \vee y = y$ and $y \wedge y = y$, therefore $L_7^\vee$ implies M_2. Besides, denoting the variables of f and g by $y_1, \ldots, y_n$ and taking $y_1 := \cdots := y_n := y$, we obtain $f(y_1, \ldots, y_n) = y = g(y_1, \ldots, y_n)$ by Corollary 1.2.1, hence (1) reduces to M_1. So Kolibiar's system M_1 is fulfilled, showing that L is a modular lattice. Now set

(2) $$y_1 \vee \ldots \vee y_n = u , \; y_1 \wedge \ldots \wedge y_n = o$$

and note that any lattice polynomial p satisfies

(3) $$o \leq p(y_1, \ldots, y_n) \leq u :$$

the easy proof is by induction on the definition of polynomials. Therefore, taking $x := z := u$ and $t := o$ in (1) we obtain $f = g$, hence $L \in$ **K**. $\square$

In the remainder of this section we are looking for self-dual equational bases which define modular varieties relative to the class **M** or to the class **L** of all lattices, using the function

(2.16) $s(x, y, z) = (x \vee (y \wedge z)) \wedge (y \vee z)$,

and the characterization (2.17) $s = \tilde{s}$ of modularity within **L**.

The following lattice identities will be needed:

(4) $s(x, y \wedge z, y \vee z) = s(x, y, z)$, $\tilde{s}(x, y \vee z, y \wedge z) = \tilde{s}(x, y, z)$.

Theorem 2.3.2. (D. Kelly and Padmanabhan [2002]) *If $p = q$ and $\tilde{p} = \tilde{q}$ define a variety **K** relative to **M**, then the self-dual identity*

(5) $s(x, y \wedge p, y \vee \tilde{q}) = \tilde{s}(x, y \vee \tilde{p}, y \wedge q)$,

*where the variables x,y do not occur in p or q, defines **K** relative to **L**.*

PROOF: If $L \in \mathbf{K}$ then (5) is readily verified. Conversely, suppose a lattice L satisfies (5), that is,

$(x \vee ((y \wedge p) \wedge (y \vee \tilde{q}))) \wedge ((y \wedge p) \vee (y \vee \tilde{q})) = (x \wedge ((y \vee \tilde{p}) \vee (y \wedge q))) \vee ((y \vee \tilde{p}) \wedge (y \wedge q))$.

Take $x := o$, $y := u$, cf. (2), where $y_1, \ldots, y_n$ are the variables of p and q. It follows by (3) that $p = q$ in L and since (5) is self-dual, it also implies that L satisfies $\tilde{p} = \tilde{q}$. Besides, taking $y_1 := \cdots := y_n := z$, identity (5) becomes $s(x, y \wedge z, y \vee z) = \tilde{s}(x, y \vee z, y \wedge z)$. Therefore, taking into account (4), we obtain (2.17), showing that $L \in \mathbf{M}$. Summarizing, we have obtained $L \in \mathbf{K}$.   □

Theorem 2.3.3. (D. Kelly and Padmanabhan [2002]) *If $p = q$ and $\tilde{p} = \tilde{q}$ define a variety **K** relative to **M**, and $p \leq q$ holds in **L**, then the self-dual identity*

(6) $s(x, p, \tilde{q}) = \tilde{s}(x, \tilde{p}, q)$,

*where the variable x does not occur in p or q, defines **K** relative to **M**. Moreover, if $p = q$ implies modularity, then (6) defines **K** relative to **L**.*

PROOF: If $L \in \mathbf{K}$ then $L \in \mathbf{M}$, hence $p = q$ and $\tilde{p} = \tilde{q}$. Using the symmetry of $s(x, y, z)$ in y, z and the characterization (2.17) of modularity, we obtain (6):

$$s(x, p, \tilde{q}) = s(x, q, \tilde{p}) = s(x, \tilde{p}, q) = \tilde{s}(x, \tilde{p}, q) .$$

Conversely, suppose a modular lattice L satisfies (6). By the same technique as in the proofs of Theorems 3.1 and 3.2, we find elements $o, u \in L$ such that $o \leq p, \tilde{p}, q, \tilde{q} \leq u$. Then from $p \leq q$ we deduce $\tilde{q} \leq \tilde{p}$ and

$$p \wedge \tilde{q} \leq p \wedge \tilde{p} \leq q \wedge \tilde{p} = s(o, q, \tilde{p}) = \tilde{s}(o, \tilde{q}, p) = p \wedge \tilde{q} ,$$

therefore $p \wedge \tilde{q} = p \wedge \tilde{p} = q \wedge \tilde{p}$. We infer similarly $p \vee \tilde{q} = p \vee \tilde{p} = q \vee \tilde{p}$, by using u instead of o. From $p \wedge \tilde{p} = q \wedge \tilde{p}$ and $p \vee \tilde{p} = q \vee \tilde{p}$ we obtain $p = q$ by Corollary 1.1. Interchanging the identities $p = q$ and $\tilde{p} = \tilde{q}$ will give a proof that $\tilde{p} = \tilde{q}$ holds in L, thereby $L \in \mathbf{K}$, completing the proof of the first statement.

Now assume that $p = q$ implies modularity and let L be a lattice satisfying (6). If we show that (6) implies modularity, it will follow from the first part of the theorem that $L \in \mathbf{K}$. So let us prove that (6) fails in any non-modular lattice. In view of Theorem 1.1, it suffices to prove this for the five-element lattice N5 depicted in Fig.1, but in which we change the notation: $x = a$, $y = b$, $z = c$.

The identity $p = q$ fails in the non-modular lattice N5 and since $p \leq q$ holds in any lattice, it follows that $p < q$ for a suitable substitution of the variables. Several cases are possible.

1. If $p = o$ then $\tilde{p} = u$ for the same substitution instance (take the upside down version of N5). Hence condition (6), which we write explicitly,

$$(6') \qquad (x \vee (p \wedge \tilde{q})) \wedge (p \vee \tilde{q}) = (x \wedge (\tilde{p} \vee q)) \vee (\tilde{p} \wedge q) ,$$

reduces to $x \wedge \tilde{q} = x \vee q$. Taking $x := o$ we see that this equality fails.

2. If $p = a$ then $\tilde{p} = c$ and there are two subcases.

2.1. If $q = c$ then $\tilde{q} = a$ and $(6')$ reduces to $(x \vee a) \wedge a = (x \wedge c) \vee c$, which is false.

2.2. If $q = u$ then $\tilde{q} = o$ and $(6')$ reduces to $x \wedge a = x \vee a$, which is false.

3. If $p = c$ then $\tilde{p} = a$ and $q = u$, hence $\tilde{q} = o$, therefore $(6')$ reduces to $x \wedge c = x \vee a$, which is false.

4. If $p = b$ then $\tilde{p} = b$ and $q = u$, hence $\tilde{q} = o$, therefore $(6')$ reduces to $x \wedge b = x \vee b$, which is false.

Thus condition (6) fails in all cases. $\qquad \square$

Corollary 2.3.1. (D. Kelly and Padmanabhan [2002]) *Every finitely-based self-dual variety of modular lattices can be defined by a single self-dual identity modulo the lattice axioms.*

PROOF: Apply Theorem 1.2.3 to the identities defining the variety $\mathbf{K}$ and to the characterization $\tilde{s} = s$ of modularity; cf.(2.17). Then the resulting equation $p = q$ characterizes $\mathbf{K}$, $p = q$ implies modularity, and since $\tilde{s} \leq s$ holds in $\mathbf{L}$, the proof of Theorem 1.2.3 immediately shows that $p \leq q$ holds in $\mathbf{L}$. Since $\mathbf{K}$ is self-dual, $\tilde{p} = \tilde{q}$ also holds in $\mathbf{K}$, so that $p = q \,\&\, \tilde{p} = \tilde{q}$ is a (redundant) definition of $\mathbf{K}$. Now the desired result follows by Theorem 3.3. $\qquad \square$

Finally let us confine to the variety of all modular lattices. We have seen in Chapter 1,§3 that the class of all lattices is characterized by the self-dual system $\mathbf{L_8}$ consisting of (possibly variants of) the two absorption laws and two "skew associativity" laws $(x \circ y) \circ z = (y \circ z) \circ x$ for $\circ = \vee, \wedge$. We have also noted in this chapter that identity (1.3) is equivalent to modularity. Therefore the self-dual system

$$\mathbf{M_8} = \mathbf{L_8} \cup \{(1.3)\}$$

characterizes modular lattices. Let us prove it is independent.

Any non-modular lattices proves the independence of (1.3). The other independence models will be based on the set $\{0, 1\}$. Taking $x \vee y = 1$ and $x \wedge y = \min(x, y)$ proves the independence of $L_7^{\vee}$. Taking $x \vee y = y$ and $x \wedge y = \min(x, y)$ proves the independence of $L_{33}^{\vee}$. The dual models prove the independence of $L_7^{\wedge}$ and $L_{33}^{\wedge}$, respectively.

Chapter 3

Distributive Lattices

The basic properties of modular and distributive lattices were discovered by Dedekind [1897] and Schröder [1890-1905]; cf. Birkhoff [1948]. Any chain is a distributive lattice; the open subsets of every topological space form a distributive lattice and so do the closed subsets. As a matter of fact, there are many examples of distributive lattices, including the important subclass of Boolean algebras; cf. next chapter.

We begin this chapter with some of the most useful characterizations of distributive lattices within the class of all lattices. Then we survey direct definitions of distributive lattices in terms of disjunction and conjunction and for bounded distributive lattices in terms of a ternary operation.

3.1. Distributive Lattices within Lattices

Distributivity is a natural condition for lattices, very much like commutativity for groups. As a matter of fact, in the early days of lattice theory, towards the end of XIX century, it was believed that all lattices were distributive. It was Schröder [1890-1905] who first proved the equivalence of conditions D_1 below and noted that calculus with groups is not distributive (cf. Anhangen 4,6). Curiously enough, Schröder could not convince Peirce; cf. Birkhoff [1948].

A lattice $(L, \vee, \wedge)$ is said to be *distributive* provided

$$D_1^\wedge \qquad x \wedge (y \vee z) = (x \wedge y) \vee (x \wedge z) .$$

It is immediately seen that condition $D_1^\wedge$ implies the identity $(2.1.1')$ defining modular lattices, therefore *any distributive lattice is modular*. The converse does not hold, as shown e.g. by the lattice in Fig.2 below, which is modular by Theorem 2.1.1, but not distributive.

An important property is that in any lattice condition $D_1^\wedge$ is equivalent to its dual

$$D_1^\vee \qquad x \vee (y \wedge z) = (x \vee y) \wedge (x \vee z) .$$

For it follows from $D_1^\wedge$ that

$$(x \vee y) \wedge (x \vee z) = ((x \vee y) \wedge x) \vee ((x \vee y) \wedge z)$$

$$= x \vee ((x \wedge z) \vee (y \wedge z)) = (x \vee (x \wedge z)) \vee (y \wedge z) = x \vee (y \wedge z) \,.$$

So $D_1^\wedge \Longrightarrow D_1^\vee$, hence $D_1^\vee \Longrightarrow D_1^\wedge$ by duality. Therefore *the principle of duality holds for distributive lattices.*

Each of the following dual inequalities characterizes distributive lattices because the converse inequalities hold in every lattice:

(1) $$(x \vee y) \wedge (x \vee z) \leq x \vee (y \wedge z) \,,$$

(1') $$x \wedge (y \vee z) \leq (x \wedge y) \vee (x \wedge z) \,.$$

Lemma 3.1.1. *If a lattice is modular but not distributive then it includes a sublattice of the form in* Fig.2.

Fig.2

PROOF: Suppose the elements x, y, z fail to satisfy $D_1^\vee$. Since

$$a \leq c \Longrightarrow a \vee (b \wedge c) = (a \vee b) \wedge c = (a \vee b) \wedge (a \vee c) \,,$$

it follows that $x \not\leq z$. Since

$$c < a \Longrightarrow a \vee (b \wedge c) = a = a \wedge (a \vee b) = (a \vee c) \wedge (a \vee b) \,,$$

it follows that $z \not\leq x$. Therefore x and z are incomparable, and similarly so are x and y. Since

$$b \leq c \Longrightarrow a \vee (b \wedge c) = a \vee b = (a \vee b) \wedge (a \vee c) \,,$$

it follows that $y \not\leq z$ and similarly $z \not\leq y$. Therefore y and z are incomparable.

Summarizing, the elements x, y, z are pairwise incomparable. Then $x \wedge y < y < y \vee z$ and setting $o = x \wedge y \wedge z$ we get $o < z < y \vee z$, because $o = z$ would imply the contradiction $z \leq x \wedge y$. Assuming $o < x \wedge y$, the elements $o, x \wedge y, y, z, y \vee z$ would be related as shown in Fig.1, which contradicts the modularity. Therefore $o = x \wedge y$ and since this conclusion has been obtained only from the pairwise incomparability of x, y, z, it follows by symmetry that $o = y \wedge z$ and $o = x \wedge z$. We deduce analogously that $x \vee y = y \vee z = x \vee z = u$. Therefore the elements o, x, y, z, u are related as shown in Fig.2. □

Theorem 3.1.1. *A lattice is distributive if and only if it includes neither a sublattice of the form in* Fig.1 (Chapter 2), *nor a sublattice of the form in* Fig.2.

Comment Loosely speaking, Theorems 2.1.1 and 1.1 characterize a structure by "forbidding" certain substructures. This reminds a famous theorem in graph theory, which characterizes the planarity of a graph by forbidding certain subgraphs. However we do not know other theorems of this type in algebra.

PROOF: If a lattice is distributive, then it is modular, hence by Theorem 2.1.1 it does not include a sublattice of the form in Fig.1, and it does not include a sublattice of the form in Fig.2 either, because the latter is not distributive.

Conversely, suppose the lattice is not distributive. If it is not modular, then by Theorem 2.1.1 it includes a sublattice as depicted in Fig.1, while in the opposite case it includes a lattice as in Fig.2 by Lemma 1.1. □

Remark 3.1.1. The lattice in Fig.1 is the five-element non-modular lattice; let us call it N5. The modular non-distributive lattice in Fig.2 has 3 pairwise incomparable elements; let us call it M3. Then Theorem 1.1 can be rephrased as follows: *A lattice is distributive if and only if it includes neither* N5, *nor* M3.

We can characterize the self dual variety of distributive lattices by a self-dual identity:

Corollary 3.1.1. *A lattice is distributive if and only if it satisfies*

$$(2) \qquad (x \wedge y) \vee (y \wedge z) \vee (x \wedge z) = (x \vee y) \wedge (y \vee z) \wedge (x \vee z) \, .$$

PROOF: If the lattice is distributive, then

$$(x \wedge y) \vee (y \wedge z) \vee (x \wedge z) = ((x \vee z) \wedge y) \vee (x \wedge z) = (((x \vee z) \wedge y) \vee x) \wedge (((x \vee z) \wedge y) \vee z)$$

$$= (x \vee z \vee x) \wedge (y \vee x) \wedge (x \vee z \vee z) \wedge (y \vee z) = (x \vee z) \wedge (x \vee y) \wedge (y \vee z) \, .$$

If the lattice is not distributive then it includes at least one of the sublattices N5, M3. But in N5

$$(x\wedge y)\vee(y\wedge z)\vee(x\wedge z) = o\vee o\vee x = x \neq (x\vee y)\wedge(y\vee z)\wedge(x\vee z) = u\wedge u\wedge z = z ,$$

while in M3

$$(x \wedge y) \vee (y \wedge z) \vee (x \wedge z) = o \neq (x \vee y) \wedge (y \vee z) \wedge (x \vee z) = u .$$

$\square$

Corollary 3.1.2. *A lattice is distributive if and only if it satisfies*

$$(3) \qquad\qquad (x \vee y) \wedge (z \vee (x \wedge y)) = (x \wedge y) \vee (y \wedge z) \vee (z \wedge x) .$$

PROOF: Identity (3) is immediate in a distributive lattice and it fails in a non-distributive lattice because in N5 and M3 we have $(x\vee y)\wedge(z\vee(x\wedge y)) = u \wedge z = z$, while in N5 we have $(x \wedge y) \vee (y \wedge z) \vee (z \wedge x) = o \vee x = x$, while in M3 we have $(x \wedge y) \vee (y \wedge z) \vee (z \wedge x) = o$. $\square$

The following analogue of Corollary 2.1.1 is valid:

Corollary 3.1.3. *A lattice is distributive if and only if*

$$(4) \qquad\qquad \exists z \, (x \wedge z = y \wedge z \,\&\, x \vee z = y \vee z) \Longrightarrow x = y .$$

PROOF: If the lattice is distributive, the left-hand side of (4) implies

$$x = x \wedge (x \vee z) = x \wedge (y \vee z) = (x \wedge y) \vee (x \wedge z)$$

$$= (x \wedge y) \vee (y \wedge z) = (x \vee z) \wedge y = (y \vee z) \wedge y = y .$$

If the lattice is not distributive, then it includes at least one of the sublattices N5, M3. In N5 we have $x \wedge y = o = z \wedge y$ and $x \vee y = u = z \vee y$ but $x \neq z$, while in M3 we have $x \wedge z = o = y \wedge z$ and $x \vee z = u = y \vee z$ but $x \neq y$. $\square$

Several modifications of condition (4) are equivalent to it:

Corollary 3.1.4. *Each of the following conditions is equivalent to the distributivity of the lattice:*

$$(5) \qquad\qquad \exists z \, (x \wedge z \leq y \wedge z \,\&\, x \vee z \leq y \vee z) \Longrightarrow x \leq y ,$$

$$(6) \qquad\qquad \exists z \, (x \wedge z \leq y \,\&\, x \leq y \vee z) \Longrightarrow x \leq y ,$$

$$(7) \qquad \exists z \, (x \wedge z \leq y \,\&\, x \leq y \vee z) \Longrightarrow \forall z \, (x \wedge z \leq y \,\&\, x \leq y \vee z) .$$

PROOF: (4)$\Longrightarrow$(5): The left-hand side of (5) implies

$$x \wedge z = x \wedge z \wedge y \wedge z = x \wedge y \wedge z \, ,$$

$$x \vee z = (x \vee z) \wedge (y \vee z) = (x \wedge y) \vee z \, ,$$

therefore $x = x \wedge y$, that is, $x \leq y$.

(5)$\Longrightarrow$(4) because $a = b \Longleftrightarrow a \leq b \ \& \ b \leq a$.

(5)$\Longleftrightarrow$(6) because their left-hand sides are clearly equivalent.

(6)$\Longrightarrow$(7) because $x \leq y$ implies the right-hand side of (7).

(7)$\Longrightarrow$(6): take $z := x$ in the right-hand side of (7). $\square$

Conditions (6) and (7) are due to Picu [1982], who provided more variations on this theme.

Theorem 1.1 suggests the question of whether one can define distributive lattices satisfying two mutually independent identities: one identity should "destroy" N5 and the other would do the same job for M3. Here is such an equational basis:

Theorem 3.1.2. *A lattice is distributive if and only if it satisfies the mutually independent axioms*

(8) $x \vee (y \wedge (x \vee z)) = (x \vee y) \wedge (x \vee z) \, ,$

(9) $x \wedge (y \vee z) = x \wedge (y \vee (x \wedge (z \vee (x \wedge y)))) \, .$

PROOF: Note that (8) is the modularity law (2.1.2). In a distributive lattice the right-hand side of (9) equals

$$x \wedge (y \vee (x \wedge z) \vee (x \wedge y)) = (x \wedge y) \vee (x \wedge z) = x \wedge (y \vee z) \, .$$

Conversely, a lattice which satisfies (8) and (9) is distributive since it does not include N5 because of (8), while M3 is also forbidden because in M3 we have $x \wedge (y \vee z) = x$ but

$$x \wedge (y \vee (x \wedge (z \vee (x \wedge y)))) = x \wedge (y \vee (x \wedge z)) = x \wedge y = o \, .$$

The axioms are independent because M3 satrisfies (8), while N5 satisfies (9):

$$x \wedge (y \vee (x \wedge (z \vee (x \wedge y)))) = x \wedge (y \vee (x \wedge z)) = x \wedge (y \vee x) = x = x \wedge (y \vee z) \, .$$

$\square$

From the axiomatic point of view, this result is interesting because it is impossible to characterize distributivity (modulo lattice theory) by any independent set with more than two identities. This is simply an equational reformulation of Theorem 1.1. Similarly, modularity cannot be characterized by any independent system with more than one identity!

Theorem 3.1.3. *If Σ is a system of mutually independent identities characterizing distributive lattices within the class of all lattices, then Σ has at most two axioms.*

PROOF: Some member of Σ fails in N5, say $f_1 = g_1$. If this axiom also fails in M3, then it already characterizes distributive lattices and since Σ is independent, it follows that $\Sigma = \{f_1 = g_1\}$. Otherwise some other member of Σ fails in M3, say $f_2 = g_2$. Then the system $\{f_1 = g_1, f_2 = g_2\}$ characterizes distributive lattices and since Σ is independent, it reduces to this system.   □

By sharp contrast, Abelian groups can be characterized modulo group theory by a set of n independent identities, for all n. In other words, the ability to characterize an equational class by demanding the non-occurrence of certain lattices is a unique feature of lattice theory and has no apparent counter-part in other algebras like groups or rings.

3.2. Defining Distributive Lattices in Terms of the Operations $\vee$ and $\wedge$

The first direct definition of distributive lattices, that is, not including a system of axioms for lattices, is the system

$$\mathbf{D}_1 = \{\mathrm{L}_1^\wedge, \mathrm{L}_2^\vee, \mathrm{L}_2^\wedge, \mathrm{L}_3^\wedge, \mathrm{L}_4^\wedge, \mathrm{D}_1^\wedge\},$$

suggested by Birkhoff [1948], Ch.IX,§1,Exercise 6 and rediscovered by Felscher [1958]. The proof is very easy: by defining $x \leq y \iff x \wedge y = x$, we obtain a meet semilattice $(L, \wedge)$. To prove that $(L, \vee, \wedge)$ is a lattice, we note that $x \leq x \vee y$ by $\mathrm{L}_4^\wedge$, $y \leq x \vee y$ by $\mathrm{L}_2^\vee$ and $\mathrm{L}_4^\wedge$, while $x \leq z$ and $y \leq z$ imply $x \vee y \leq z$ because

$$(x \vee y) \wedge z = z \wedge (x \vee y) = (z \wedge x) \vee (z \wedge y) = (x \wedge z) \vee (y \wedge z) = x \vee y.$$

Ponticopoulos [1962] provided two 3-identity bases involving 12 variables each. The 9-axiom system devised by Ellis [1949] uses D_1, L_2, L_3, a variant of L_1 and (1.2.1).

Two other systems were given by Rudeanu and Vaida [2004]:

$$\mathbf{D}_2 = \{\mathrm{L}_1^\wedge, \mathrm{L}_2^\vee, \mathrm{L}_2^\wedge, \mathrm{L}_4^\vee, \mathrm{D}_1^\wedge\},$$

$$\mathbf{D}_3 = \{L_1^\wedge, L_4^\vee, L_5^\vee, L_6^\vee, L_7^\vee, D_1^\wedge\} \, .$$

In view of McKenzie's Theorem 1.3.3, the variety of distributive lattices cannot be defined by a single identity. The following two-identity base has been widely used in the literature:

Theorem 3.2.1. (Sholander [1951]) *An algebra* $(L, \vee, \wedge)$ *of type* (2,2) *is a distributive lattice if and only if it satisfies the system*

$$\mathbf{D}_4 = \{L_4^\wedge, D_2^\wedge\} \, ,$$

where

$L_4^\wedge \qquad x \wedge (x \vee y) = x \, ,$

$D_2^\wedge \qquad x \wedge (y \vee z) = (z \wedge x) \vee (y \wedge x) \, ,$

Comment See Appendix A for a proof provided by the computer program Prover9.

PROOF: Necessity is trivial. Conversely, suppose L satisfies $\mathbf{D}_4$. We first prove the idempotency laws and the commutativity of $\wedge$. The hypothesis implies

(1) $\qquad x = x \wedge (x \vee x) = (x \wedge x) \vee (x \wedge x) \, ;$

then from (1) and $L_4^\wedge$ we obtain

(2) $\qquad x \wedge x = (x \wedge x) \wedge ((x \wedge x) \vee (x \wedge x)) = (x \wedge x) \wedge x$

and using (1), $D_2^\wedge$, (2) and (1) we deduce

$x \wedge x = x \wedge ((x \wedge x) \vee (x \wedge x)) = ((x \wedge x) \wedge x) \vee ((x \wedge x) \wedge x) = (x \wedge x) \vee (x \wedge x) = x \, ,$

so that (1) becomes $x = x \vee x$. This yields further, via $D_2^\wedge$,

$$x \wedge y = (x \wedge y) \vee (x \wedge y) = y \wedge (x \vee x) = y \wedge x \, .$$

Now we use freely idempotency, the commutativity of $\wedge$ and $\mathbf{D}_4$ to obtain the commutativity of $\vee$. We have in turn

(3) $\qquad x = x \wedge (x \vee y) = (y \wedge x) \vee (x \wedge x) = (y \wedge x) \vee x \, ,$

(4) $\qquad \begin{aligned} x &= x \wedge x = x \wedge ((y \wedge x) \vee x) \\ &= (x \wedge x) \vee ((y \wedge x) \wedge x) = x \vee ((y \wedge x) \wedge x) \, , \end{aligned}$

(5) $\qquad \begin{aligned} x \vee y &= (x \vee y) \wedge (x \vee y) = (y \wedge (x \vee y)) \vee (x \wedge (x \vee y)) \\ &= ((y \wedge y) \vee (x \wedge y)) \vee x = (y \vee (x \wedge y)) \vee x \, , \end{aligned}$

and it follows from (3), (4) and (5) that

(6) $y = (x \wedge y) \vee y = (y \vee ((x \wedge y) \wedge y)) \vee (x \wedge y) = y \vee (x \wedge y)$;

from (4) and (6) we get $x \vee y = y \vee x$.

We have thus established properties L_1, L_2 and L_4. We prepare the proof of $L_3^\vee$ by computing

$$c \wedge ((a \vee b) \vee c) = (c \wedge (a \vee b)) \vee (c \wedge c) = (c \wedge (a \vee b)) \vee c = c \,,$$

$$b \wedge ((a \vee b) \vee c) = (b \wedge (a \vee b)) \vee (b \wedge c) = b \vee (b \wedge c) = b \,,$$

and similarly $a \wedge ((a \vee b) \vee c) = a$. Set $(a \vee b) \vee c = P$ and $a \vee (b \vee c) = Q$. Then, using $D_2^\wedge$, we obtain

$$Q = (a \wedge P) \vee ((b \wedge P) \vee (c \wedge P)) = (a \wedge P) \vee ((b \vee c) \wedge P) = (a \vee (b \vee c)) \wedge P = Q \wedge P$$

and quite similarly we deduce $P = P \wedge Q$, therefore $P = Q$, that is, $L_3^\vee$.

In the above proof of $L_3^\vee$ we have used the distributivity $D_2^\wedge$, which reduces to $D_1^\wedge$. Note that the proof of the implication $D_1^\wedge \implies D_1^\vee$, given in the beginning of this chapter, uses in fact $D_1^\wedge, L_2, L_4$ and $L_3^\vee$, therefore D_4 implies also $D_1^\vee$. It follows that the proof of $L_3^\vee$ can be dualized and we obtain $L_3^\wedge$ as well.   □

We now pass to direct definitions of distributive lattices with 0 or with 1 or with 0 and 1, that is, we include 0 and/or 1 in the type of the algebra, even in the cases when the original papers work with algebras of type (2,2); see e.g. the next system.

The first such system of axioms, namely

$$\mathbf{D}_1^1 = \{L_3^\wedge, D_1^\wedge, D_3^\wedge, L_{\vee 1}, L_{1\vee}, L_{\wedge 1}, L_{1\wedge}\} \,,$$

where we have set

$D_3^\wedge$ $(y \vee z) \wedge x = (y \wedge x) \vee (z \wedge x)$,

$L_{\vee 1}$ $x \vee 1 = 1$, $L_{1\vee}$ $1 \vee x = 1$,

$L_{\wedge 1}$ $x \wedge 1 = x$, $L_{1\wedge}$ $1 \wedge x = x$,

defines distributive lattices with greatest element 1 and is essentially due to G.D.Birkhoff and G. Birkhoff [1946]; see also Birkhoff [1948], Ch.IX, Theorem 3. The problem of the independence of this system, known as Birkhoff's Problem 65, was solved in the affirmative by Croisot [1951], Wooyenaka [1951], Matusima [1952], Szász [1952] and Zelinka [1967], independently of each other.

As a matter of fact, in the original paper axioms $L_{\vee 1}, \ldots, L_{1\wedge}$ appear in the form "there is 1 such that $x \vee 1 = 1$ for all x" etc. However, the authors implicitly admit that the element 1 is the same in all the axioms. As explained above, we disregard this formulation and interpret $\mathbf{D}_1^1$ as referring to an algebra $(L, \vee, \wedge, 0, 1)$ of type (2,2,0,0); so did Matusima, Szász and Zelinka. In his paper, Croisot puts the questionable axioms in the form

$$\exists 1 \; x \vee 1 = 1 \,, \qquad \exists 1' \; 1' \vee x = 1' \,,$$

$$\exists 1'' \; x \wedge 1' = x \,, \qquad \exists 1''' \; 1''' \wedge x = x \,,$$

adds the axiom $1 = 1''$, states that one can prove $1 = 1'$ and $1'' = 1'''$, and proves the independence of the new system. He then proves that axioms $1' \vee x = 1, 1''' \wedge x = x, D_1^\wedge$ and $D_3^\wedge$ can be replaced by $D_2^\wedge$,[†] and this reduced system is also independent. Wooyenaka provided four independent systems by a similar approach.

Sobociński [1972a,b] provided a variant of the Croisot system, then simplified it as follows:

$$\mathbf{D}_2^1 = \{D_4^\wedge, L_{\vee 1}, L_{\wedge 1}\} \,,$$

where

$$D_4^\wedge \qquad x \wedge ((y \wedge y) \vee z) = (z \wedge x) \vee (y \wedge x) \,.$$

The simplest direct definition of distributive lattices with 1 is the dual of a theorem which characterizes distributive lattices with 0 and was found by Ferentinou-Nicolacopoulou [1969] and Tamura [1975], independently of each other:

Theorem 3.2.2. *Each of the following systems of axioms characterizes distributive lattices with greatest element within algebras* $(L, \vee, \wedge, 1)$ *of type* (2,2,0):

$$\mathbf{D}_3^1 = \{D_5^\vee, L_4^\vee\} \,,$$

$$\mathbf{D}_4^1 = \{D_6^\vee, L_4^\vee\} \,,$$

where

$$D_5^\vee \qquad x \vee (y \wedge z) = (z \vee (x \wedge 1)) \wedge (y \vee (x \wedge 1)) \,,$$

$$D_6^\vee \qquad x \vee (y \wedge z) = (z \vee (x \wedge 1)) \vee (y \wedge x) \,.$$

PROOF: We begin with $\mathbf{D}_3^1$. Necessity is trivial. Conversely, suppose L satisfies $\mathbf{D}_2^1$. Then we obtain in turn

$$x \vee (x \wedge 1) = x \,,$$

[†]This axiom appears simultaneously in the papers by Croisot and Sholander.

$$x = x \vee (x \wedge x) = (x \vee (x \wedge 1)) \wedge (x \vee (x \wedge 1)) = x \wedge x ,$$

$$x = x \vee (x \wedge x) = x \vee x ,$$

$$x \vee y = x \vee (y \wedge y) = (y \vee (x \wedge 1)) \wedge (y \vee (x \wedge 1)) = y \vee (x \wedge 1) ,$$

and from the latter identity with $y := x$ then with $y := x \wedge 1$, we get

$$x = x \vee x = x \vee (x \wedge 1) = (x \wedge 1) \vee (x \wedge 1) = x \wedge 1 ,$$

hence $D_5^\vee$ reduces to the dual $D_2^\vee$ of $D_2^\wedge$, therefore $(L, \vee, \wedge)$ is a distributive lattice by the dual of Theorem 2.1 and it satisfies $x = x \wedge 1$ as well.

System $\mathbf{D}_4^1$ is left to the reader. $\square$

Another 2-axiom system for distributive lattices with 1 was given by Tamura [1975].

Iséki and Ôhashi [1970] presented (without proofs) four equational bases for bounded distributive lattices, each of them having 5 axioms and involving 6,6,7 and 8 variables, respectively. One of these systems corrects an error in Ôhashi [1968], pointed out by Lowig [1969]. A two-axiom system involving 8 variables was provided by Sobociński [1979].

Note that a short definition of bounded distributive lattices can be obtained by adding one axiom to a short definition of distributive lattices with 1/with 0, or else, according to a remark of Sholander [1951], by adding the axiom $0 \vee (x \wedge 1) = x$ or its dual to a short definition of distributive lattices. For this axiom can be written in the form $(0 \vee x) \wedge (0 \vee 1) = x$, hence $x \leq 0 \vee 1$, showing that $0 \vee 1$ is the greatest element, so that the latter identity reduces to $0 \vee x = x$, therefore the original axiom becomes $x \wedge 1 = x$.

3.3. Defining Bounded Distributive Lattices in Terms of the Median Operation

Recall that a latice with 0 and 1 is said to be *bounded*. This section is devoted to various characterizations of bounded distributive lattices in terms of the *median operation*

$$(1) \qquad\qquad m(x, y, z) = (y \wedge z) \vee (x \wedge z) \vee (x \wedge y) ,$$

which was already given as an example of a majority polynomial in arbitrary lattices.

It was seen in Corollary 1.1 that distributive lattices are characterized by the identity $m = \tilde{m}$, so that we have also

(2)
$$m(x, y, z) = (y \vee z) \wedge (x \vee z) \wedge (x \vee y) \, .$$

As a matter of fact

$$m(x, y, z) = (y \wedge z) \vee (x \wedge (y \vee z)) = \tilde{s}(x, y, z) \, ,$$

where s is the function studied in Chapter 2, where it was seen that modular lattices are characterized by the identity $s = \tilde{s}$.

In a bounded distributive lattice identities (2) and (1) imply

(3)
$$x \vee y = m(x, 1, y) \, , \qquad x \wedge y = m(x, 0, y) \, .$$

The results of this section fall under the following general scheme. A set P of polynomial identities of type (3,0,0) is said to *characterize* bounded distributive lattices if the following hold: α) the median operation (1) of every bounded distributive lattice $(L, \vee, \wedge, 0, 1)$ satisfies P, and β) conversely, if an algebra $(L, m, 0, 1)$ of type (3,0,0) satisfies P, then the algebra $(L, \vee, \wedge, 0, 1)$ defined by (3) is a bounded distributive lattice in which (1) holds.

From a practical point of view, the above point α) is a matter of routine computation which raises no difficulty, so that the proofs reduce to establishing β).

Lemma 3.3.1. *If a ternary operation τ satisfies*

$$\tau(x, y, z) = \tau(y, x, z) \text{ and } \tau(x, y, z) = \tau(x, z, y) \, ,$$

then τ is invariant under any permutation of the variables.

PROOF: Follows from

$$\tau(x, y, z) = \tau(y, x, z) = \tau(y, z, x) = \tau(z, y, x) = \tau(z, x, y) = \tau(x, z, y) \, .$$

$\square$

Theorem 3.3.1. (Birkhoff and Kiss [1947]) *Bounded distributive lattices are characterized by the system*

$$\mathbf{D}_1^{01} = \{D_6, D_7, D_8, D_9, D_{10}\} \, ,$$

where

D_6 $m(x, y, z) = m(y, x, z) \, ,$

D_7 $m(x, y, z) = m(x, z, y) \, ,$

D_8 $m(x, y, x) = x \, ,$

D_9 $m(0, x, 1) = x \, ,$

D_{10} $m(m(x, y, z), t, v) = m(m(x, t, v), y, m(z, t, v)) \, .$

PROOF: It follows from D_6 and D_7 via Lemma 3.1 that m is invariant under any permutation of the variables, therefore the operations $\vee$ and $\wedge$ defined by (3) are commutative, hence in order to check first system $\mathbf{D}_1^1$ it suffices to prove $L_3^{\wedge}, L_{\vee 1}, L_{\wedge 1}, D_1^{\wedge}$:

$$x \wedge x = m(x, 0, x) = x \ ,$$

$$x \vee 1 = m(x, 1, 1) = m(1, x, 1) = 1 \ ,$$

$$x \wedge 1 = m(x, 0, 1) = m(0, x, 1) = x \ ,$$

$$x \wedge (y \vee z) = m(x, 0, m(y, 1, z)) = m(m(y, 1, z), 0, x)$$

$$= m(m(y, 0, x), 1, m(z, 0, x)) = m(x \wedge y, 1, x \wedge z) = (x \wedge y) \vee (x \wedge z) \ ,$$

$$x \wedge 0 = m(x, 0, 0) = m(0, x, 0) = 0 \ .$$

The last point is to prove identity (1). Indeed, by applying in turn (3), (3), symmetry, D_{10}, symmetry, D_8 and (3), we obtain

$$(x \wedge y) \wedge m(x, y, z) = m(x \wedge y, 0, m(x, y, z)) = m(m(x, 0, y), 0, m(x, y, z))$$
$$= m(m(0, x, y), 0, m(z, x, y)) = m(m(0, 0, z), x, y)$$
$$= m(m(0, z, 0), x, y) = m(0, x, y) = x \wedge y \ ,$$

hence $x \wedge y \leq m(x, y, z)$ and similarly $y \wedge z \leq m(x, y, z)$ and $z \wedge x \leq m(x, y, z)$, therefore

$$(x \wedge y) \vee (y \wedge z) \vee (z \wedge x) \leq m(x, y, z)$$

and a similar proof shows that

$$m(x, y, z) \leq (x \vee y) \wedge (y \vee z) \wedge (z \vee x) \ ,$$

whence identity (1) follows by Corollary 3.1.1.   □

The independence of the above system was not studied. Cheremisin [1958] claimed that axiom D_7 follows from the others, but this is wrong. However Birkhoff [1948] suggested that one of the axioms D_6 and D_7 or both can be dispensed with if axiom D_{10} is suitably modified; this is known as Birkhoff's Problem 64. Several solutiona have been proposed.

The first solutions to Problem 64 were given by Vassiliou [1950]*, the simplest one being

$$\mathbf{D}_2^{01} = \{D_8, D_9, D_{11}\} \ ,$$

where

D_{11} $m(x, m(y, z, t), v) = m(m(v, x, t), z, m(v, x, y))$.

Several authors found other solutions: Croisot [1951], who gave the independent system

$$\mathbf{D}_3^{01} = \{D_8, D_9, D_{12}\} \, ,$$

where

D_{12} $m(x, m(y, z, t), v) = m(z, m(t, x, v), m(y, x, v))$,

Hashimoto [1951], who devised the systems

$$\mathbf{D}_4^{01} = \{D_8, D_9, D_{13}\} \, ,$$

$$\mathbf{D}_5^{01} = \{D_8, D_9, D_{14}\} \, ,$$

where

D_{13} $m(x, m(y, z, t), v) = m(m(v, z, x), y, m(v, t, x))$,

D_{14} $m(m(x, y, z), t, m(x, v, z)) = m(m(z, t, x), y, m(z, v, x))$,

Trevisan [1951], who suggested the basis

$$\mathbf{D}_6^{01} = \{D_6, D_8, D_9, D_{15}\} \, ,$$

where

D_{15} $m(m(x, y, z), t, v) = m(m(x, v, t), y, m(z, t, v))$,

Sholander [1951], [1952], who proposed the systems

$$\mathbf{D}_7^{01} = \{D_{16}, D_{17}\} \, ,$$

$$\mathbf{D}_8^{01} = \{D_9, D_{18}, D_{19}\} \, ,$$

($\mathbf{D}_7^{01}$ without proof), where

D_{16} $m(0, x, m(1, y, 1)) = x$,

D_{17} $m(x, m(y, z, t), v) = m(m(x, z, v), t, m(y, x, v))$,

D_{18} $m(x, x, y) = x$,

D_{19} $m(m(x, y, z), t, v) = m(m(t, v, x), m(t, v, y), z)$,

Wang [1953], who proposed the basis

$$\mathbf{D}_9^{01} = \{D_8, D_9, D_{20}\} \, ,$$

where

D_{20} $m(x, m(y, z, t), v) = m(m(v, z, x), m(x, y, v), t)$, and Sobociński [1962], who provided six independent systems, namely

$$\mathbf{D}_{10}^{01} = \{D_8, D_9, D_{21}\} \,,$$

$$\mathbf{D}_{11}^{01} = \{D_{22}, D_{23}\} \,,$$

$$\mathbf{D}_{12}^{01} = \{D_{21}, D_{24}\} \,,$$

$$\mathbf{D}_{13}^{01} = \{D_{21}, D_{25}\} \,,$$

$$\mathbf{D}_{14}^{01} = \{D_8, D_{26}\} \,,$$

$$\mathbf{D}_{15}^{01} = \{D_8, D_{27}\} \,,$$

where

D_{21} $\quad m(x, m(y, z, t), v) = m(m(x, t, v), m(x, y, v), z) \,,$

D_{22} $\quad m(0, m(y, x, x), 1) = x \,,$

D_{23} $\quad m(x, m(y, z, t), v) = m(m(x, t, v), m(x, y, v), z) \,,$

D_{24} $\quad m(0, m(x, x, y), 1) = x \,,$

D_{25} $\quad m(0, m(x, y, x), 1) = x \,,$

D_{26} $\quad m(0, m(x, m(y, z, t), v), 1) = m(m(x, t, v), m(x, y, v), z) \,,$

D_{27} $\quad m(x, m(y, z, t), v) = m(0, m(m(x, t, v), m(x, y, v), z), 1) \,.$

Note that each of the systems $\mathbf{D}_1^{01} - \mathbf{D}_{15}^{01}$ involves 5 variables. Four systems making use of only 4 variables were given by authors who, curiously enough, do not mention Problem 64. Namely, Kolibiar and Marcisová [1974] proposed the system

$$\mathbf{D}_{16}^{01} = \{D_6, D_7, D_{28}, D_{29}\} \,,$$

where

D_{28} $\quad m(x, y, y) = y \,,$

D_{29} $\quad m(m(x, y, z), t, z) = m(x, z, m(t, z, y)) \,,$

while Zaĭchik [1974] devised the systems

$$\mathbf{D}_{17}^{01} = \{D_6, D_7, D_8, D_9, D_{30}\} \,,$$

$$\mathbf{D}_{18}^{01} = \{D_8, D_9, D_{31}\} \,,$$

$$\mathbf{D}_{19}^{01} = \{D_{25}, D_{31}\} \,,$$

where

D_{30} $\quad m(m(x,y,z),t,x) = m(x,y,m(z,t,x))$,

D_{31} $\quad m(x,m(y,z,t),y) = m(y,m(z,x,y),t)$.

Malliah [1979] announced without proofs many independent systems which solve Problem 64: 313 systems of 3 axioms, 9 bases having 4 axioms and several systems containing 5 axioms, obtained by permutations of variables. The author claimed that these systems includes all previously known solutions to Problem 64, but the references are incomplete. As an example, she proved that the system

$$\mathbf{D}_{20}^{01} = \{D_8, D_9, D_{32}\} ,$$

where

D_{32} $\quad m(m(x,y,z),t,v) = m(m(v,t,x),y,m(v,t,z))$,

is independent and characterizes bounded distributive lattices. Finally the author provided with full proofs the following 15 independent two-identity systems:

D_{33} $\quad m(x,m(0,y,1),x) = x$,

together with one of

D_{34} $\quad m(0,m(x,m(y,z,t),v),1) = m(m(x,z,v),t,m(x,y,v))$,

D_{35} $\quad m(x,m(y,z,t),v) = m(0,m(m(x,z,v),t,m(x,y,v)),1)$,

D_{36} $\quad m(m(0,x,1),m(y,z,t),m(0,v,1)) = m(m(x,z,v),t,m(x,y,v))$,

D_{37} $\quad m(x,m(y,z,t),v) = m(m(0,m(x,z,v),1),t,m(0,m(x,y,v),1))$,

D_{38} $\quad m(m(0,x,1),m(0,m(y,z,t),1),m(0,v,1))$
$\qquad = m(m(x,z,v),t,m(x,y,v))$,

D_{39} $\quad m(m(y,z,t),v)$
$\qquad = m(m(0,m(x,z,v),1),m(0,t,1),m(0,m(x,y,v),1))$;

then D_8 together with one of

D_{40} $\quad m(0,m(x,m(y,z,t),v),1) = m(m(x,z,t),t,m(x,y,v))$,

D_{41} $\quad m(m(0,x,1),m(y,z,t),m(0,v,1)) = m(m(x,z,v),z,m(x,y,v))$,

D_{42} $\quad m(m(0,x,1),m(0,m(y,z,t),1),m(0,v,1))$
$\qquad = m(m(x,z,v),t,m(x,y,v))$,

D_{43} $\quad m(x,m(y,z,t),v) = m(0,m(x,z,v),t,m(x,y,v),1)$,

D_{44} $\quad m(x,m(y,z,t),v) = m(m(0,m(x,z,v),1),t,m(0,m(x,y,v),1))$,

D_{45} $\quad m(x,m(y,z,t),v)$
$\qquad = m(m(0,m(x,z,v),1),m(0,t,1),m(0,m(x,y,v),1))$;

finally

D_{46} $m(x, m(y, z, t), v) = m(m(x, z, v), t, m(x, y, v))$,

together with one of

D_{47} $m(0, m(x, y, x), 1) = x$,

D_{48} $m(0, m(y, x, x), 1) = x$,

D_{49} $m(0, m(x, x, y), 1) = x$.

Drašičková [1966]*, working within algebras $(L, m, 0)$ of type $(3,0)$, proposed a variant of system $\mathbf{D}_1^{01}$ in which the invariance of m with respect to permutations is postulated and D_9 is replaced by the weaker axiom

$$\forall x \, \forall y \, \exists u \, m(0, x, u) = x \ \& \ m(0, y, u) = y \ .$$

Sholander [1952] defined bounded distributive lattices in terms of the segments already mentioned in Chapter 1:

(1.4.1) $[a, b] = \{x \in L \mid a \wedge b \leq x \leq a \vee b\}$.

Pic [1969] made an attempt to obtain characterizations of bounded distributive lattices in terms of an n-ary operation instead of a ternary one.

Chapter 4

Boolean Algebras

Boolean algebras have a special position in lattice theory. The concept of Boolean algebra and the more general concept of a lattice crystallized more or less simultaneously, towards the end of the XIX-th century; cf. historical notes in Birkhoff [1948]. Ever since then the theory of Boolean algebras has tremendously developed, as well as its numerous applications in logic, measure theory, probability theory, topology, computer science and others. In particular there are more essentially distinct sets of axioms for Boolean algebras than for any other class of lattices.

One of the factors which produce the diversity of the axiomatics of Boolean algebras is the choice of the signature. In this chapter we basically regard a Boolean algebra as an algebra $(B, \vee, \wedge,' , 0, 1)$ of type $(2,2,1,0,0)$, but other definitions work with shorter types. Thus, a Boolean algebra is uniquely determined by its bounded-lattice reduct $(B, \vee, \wedge, 0, 1)$ and even by its lattice reduct $(B, \vee, \wedge)$; the reduct $(B, \vee,')$ determines the whole Boolean-algebra structure as well. A Boolean algebra is also equivalent with each of the algebras $(B, -, 1)$, $(B, |)$, where $x - y = x \wedge y'$ and $x|y = x' \wedge y'$, with $(B, m,')$, where m is the median operation, and with a Boolean ring. These equivalences fall under the following general scheme.

Let Σ and Θ be two signatures. A class $\mathcal{A}$ of $\Sigma-$algebras is *definitionally equivalent* to a class $\mathcal{B}$ of $\Theta-$algebras provided there is a family Φ of $\Sigma-$polynomials and a family Ψ of $\Theta-$polynomials such that for every $(B, F) \in \mathcal{A}$ and every $(B, G) \in \mathcal{B}$ the following hold: (i) $(B, \Phi(F)) \in \mathcal{B}$, (ii) $(B, \Psi(G)) \in \mathcal{A}$, and (iii) $\Psi\Phi(F) = F$ and $\Phi\Psi(G) = G$. It follows that the categories $\mathcal{A}$ and $\mathcal{B}$ are isomorphic. In particular the relationships between a reduct and an extension mentioned above are obtained for: $\Theta \subset \Sigma$, $G \subset F$, then $\Phi = (\varphi_g)_{g \in \Theta}$ with $\varphi_g(F) = g$ for all $g \in \Theta$, and $\Psi = (\psi_f)_{f \in \Sigma}$ with $\psi_f(G) = f$ for all $f \in \Sigma$.

We will apply the above scheme to the class $\mathcal{A}$ of Boolean algebras. Each system of axioms defines a class $\mathcal{B}$ of $\Theta-$algebras and it turns out that properties (i) and $\Psi\Phi = \mathbf{1}$ are immediate, so that the proof reduces to establishing that Ψ transforms a Θ-algebra into a Boolean algebra and $\Phi\Psi = 1$, which is done by using the system of axioms.

Note that definitional equivalence is stronger than the equivalence between two axiom systems of different signatures, usually met in the literature. The latter concept requires only properties (i) and (ii). Yet property $\Psi\Phi = 1$ is easily checked in most cases.

The present chapter is organized as follows. The first three sections present Boolean algebras within lattices. In §1 we list systems of axioms in which complementation is not taken as one of the basic operations; instead, an axiom states that every element has (at least) a complement and the uniqueness of the complement must be proved. In the next two sections complementation is taken as a basic operation. So §2 deals with Boolean algebras expressed in terms of join, meet and complementation, while §3 presents systems of axioms in terms of join and complementation only, which is possible due to the De Morgan law. The axiomatics of the important concept of Boolean ring is the subject matter of §4. Thus in §§2–4 Boolean algebra are defined by associative operations, while in §5 the definitions use nonassociative binary operations: difference or implication or Sheffer stroke. As shown in §6, Boolean algebras can also be defined by using ternary operations: majority decision m, or rejection m', or in terms of n-ary Sheffer functions. The tools used in §7 are the partial order $\leq$ and the relation $x\chi y \iff x \wedge y = 0$. In §8 we present several characterizations of Boolean algebras within uniquely complemented lattices; note that §§1,8 are an exception to the policy announced in the Introduction, namely to omit characterizations of a class of lattices within a larger class. So this chapter illustrates the fact that Boolean algebras can be presented as algebras of various signatures; the metatheorem proved in §9 says that Boolean algebras can be characterized by a single identity, whatever be the type. On the other hand, Boolean algebras are related to orthomodular lattices both by their origins (classical propositional calculus and quantum logic, respectively) and by the interferences between their axiomatics. That is why we have included a final section dealing with orthomodular lattices.

In view of a uniform approach, the following condition is understood throughout this chapter:

every system of axioms requires that the support set B of the algebra be of cardinality at least 2,

even if the original paper does not mention this. Whenever the type of the algebra presupposes two constants 0 and 1, this tacit condition has the form $0 \neq 1$.

We also apply a uniform treatment to the constants 0 and/or 1 : unless otherwise stated, we include them among the basic operations, even if the original formulation was something like "there is an element 0 such that $x \vee 0 = x$ for all x". As a matter of fact, we already did so in the previous chapters.

4.1. Boolean Lattices

Boolean lattices are defined as complemented distributive lattices; they are usually known as Boolean algebras, implying that complementation is incorporated as a basic operation, along with join, meet, 0 and 1.

Let L be a bounded lattice. We regard it as an algebra $(L, \vee, \wedge, 0, 1)$ of type or *signature* (2,2,0,0). We say that an element $x \in L$ is *complemented* if there is an element $y \in L$ such that $x \vee y = 1$ and $x \wedge y = 0$; then y is called a *complement* of x. Note that 0 and 1 are complements of each other. We denote by $C(L)$ the set of complemented elements of L.

Proposition 4.1.1. *In a bounded distributive lattice an element can have at most one complement.*

PROOF: Suppose y and z are complements of x. Then

$$y = y \wedge 1 = y \wedge (x \vee z) = (y \wedge x) \vee (y \wedge z) = 0 \vee (y \wedge z)$$

$$= (x \wedge z) \vee (y \wedge z) = (x \vee y) \wedge z = 1 \wedge z = z \ .$$

$\square$

We will use the notation x' for the possible unique complement of an element x of a bounded distributive lattice.

Proposition 4.1.2. *If L is a bounded distributive lattice, then $C(L)$ is a subalgebra of the algebra $(L, \vee, \wedge, 0, 1)$ and for every $x, y \in L$ we have*

(1) $$(x \vee y)' = x' \wedge y' \ ,$$

(2) $$(x \wedge y)' = x' \vee y' \ ,$$

(3) $$x'' = x \ .$$

Comment Properties (1), (2) are known as the *De Morgan laws*, while (3) is called the *law of double negation*.

PROOF: We use Proposition 1.1. Property (3) is just a restatement of the identities $x' \vee x = 1$ and $x' \wedge x = 0$. Property (2) is the dual of (1), and the latter follows from

$$(x \vee y) \vee (x' \wedge y') = (x \vee y \vee x') \wedge (x \vee y \vee y') = 1 \wedge 1 = 1 \,,$$

$$(x \vee y) \wedge (x' \wedge y') = (x \wedge x' \wedge y') \vee (y \wedge x' \wedge y') = 0 \vee 0 = 0 \,.$$

$\square$

Note that a bounded lattice such that $0 = 1$ is *trivial*, that is, it reduces to a singleton. For $x = x \wedge 1 = x \wedge 0 = 0$.

By a *complemented lattice* we mean a bounded lattice L such that all of its elements are complemented, that is, such that $L = C(L)$. A non-trivial complemented distributive lattice will be called a *Boolean lattice*. Applying again Proposition 1.1, we see that a Boolean lattice can be endowed with a unary operation $'$ called *complementation*, which sends each element x to its complement x'. This enriched structure is called a Boolean algebra. In other words, a *Boolean algebra* is an algebra $(B, \vee, \wedge, ', 0, 1)$ of type $(2,2,1,0,0)$ such that its reduct $(B, \vee, \wedge, 0, 1)$ is a Boolean lattice and the unary operation $'$ is complementation.

Remark 4.1.1. As a matter of fact, in the definition of complemented lattices (not necessarily distributive) it suffices to suppose that 0 and 1 are two distinguished elements of L, and it will follow that they are bounds of L. For $x \vee y = 1$ implies $x \leq 1$, while from $x \wedge y = 0$ we infer $0 \leq x$, and these inequalities hold for all x.

We have thus presented the concept of Boolean algebra as it stands nowadays. However this axiomatic way of thinking has emerged from a more "computational" point of view of Boole himself and his direct followers, who were interested in computing with the truth values 0 and 1, or equivalently, with sets ("classes", "Gebiete"); see e.g. Boole [1847], [1854], Schröder [1890-1905]. Here are the most important properties they have found:

$B_1^\vee$ $\quad$ $x \vee x = x \,,$

$B_1^\wedge$ $\quad$ $x \wedge x = x \,,$

$B_2^\vee$ $\quad$ $x \vee y = y \vee x \,,$

$B_2^\wedge$ $\quad$ $x \wedge y = y \wedge x \,,$

$B_3^\vee$ $\quad$ $(x \vee y) \vee z = x \vee (y \vee z) \,,$

$B_3^\wedge$ $\quad$ $(x \wedge y) \wedge z = x \wedge (y \wedge z) \,,$

$B_4^\vee$ $\quad$ $x \vee (x \wedge y) = x \,,$

$B_4^\wedge$ $x \wedge (x \vee y) = x$,

$B_5^\vee$ $x \vee (y \wedge z) = (x \vee y) \wedge (x \vee z)$,

$B_5^\wedge$ $x \wedge (y \vee z) = (x \wedge y) \vee (x \wedge z)$,

$B_6^\vee$ $x \vee 0 = x$,

$B_6^\wedge$ $x \wedge 1 = x$,

$B_7^\vee$ $x \vee 1 = 1$,

$B_7^\wedge$ $x \wedge 0 = 0$,

$B_8^\vee$ $x \vee x' = 1$,

$B_8^\wedge$ $x \wedge x' = 0$,

B_9 $x'' = x$.

From the contemporary point of view, identities $B_1 - B_9$ can be regarded as a highly redundant system of axioms for Boolean algebras, while the compact definition of Boolean lattices as complemented distributive lattices is due to Birkhoff [1948].

In this section we will present systems of axioms for Boolean lattices. Note that any such system must contain (at least) an axiom which is not an identity but states that for each element x there is an element x' satisfying properties B_8 or some other property equivalent to B_8. Unless otherwise stated, these systems refer to algebras $(B, \vee, \wedge, 0, 1)$ of type $(2,2,0,0)$.

The first system of axioms for Boolean lattices and for Boolean algebras in general was given by Whitehead [1898]. It is worth mentioning that his Treatise on Universal Algebra is indeed a book on universal algebra in the modern sense of the word! In particular, at that time Boolean algebra was "the only known member of the non-numerical genus of universal algebra" as noted by Whitehead; cf. Birkhoff [1948], footnote 1 to Ch.X. So Whitehead may be viewed as the founder of the modern concept of Boolean algebra.

The system given by Whitehead is

$$\mathbf{B}_1 = \{B_1^\vee, B_1^\wedge, B_2^\vee, B_2^\wedge, B_3^\vee, B_3^\wedge, B_4^\vee, B_5^\wedge, B_6^\wedge, B_{10}\} ,$$

where

B_{10} $\forall x \; \exists x' \; x \vee x' = 1 \; \& \; x \wedge x' = 0$.

To see that the algebra defined by system $\mathbf{B}_1$ is actually a Boolean lattice, it remains to prove that it is a distributive lattice. This follows from Theorem 3.2.1, because the commutativity B_2 enables one to write axiom $B_5^\wedge$ in the form $D_2^\wedge$, while $B_5^\wedge, B_1^\wedge$ and $B_4^\vee$ imply

$$x \wedge (x \vee y) = (x \wedge x) \vee (x \wedge y) = x ,$$

that is, $B_4^\wedge$, which coincides with $L_4^\wedge$. Incidentally, we have also proved that axioms $B_1^\vee, B_3^\vee$ and $B_3^\wedge$ are redundant.

Huntington [1904], [1932], [1933a], [1933b] proposed several systems of axioms characterizing Boolean algebras; those given in [1904], [1933a] are known as *Huntington's first-sixth sets of postulates* for Boolean algebras. *Huntington's first set* is essentially

$$\mathbf{B}_2 = \mathbf{I} = \{B_2^\vee, B_2^\wedge, B_5^\vee, B_5^\wedge, B_6^\vee, B_6^\wedge, B_{10}\} \ .$$

To prove the correctness of system $\mathbf{B}_2$, it remains to show that every algebra satisfying this system is a distributive lattice. In view of Sholander's Theorem 3.2.1, it suffices to prove that $\mathbf{B}_2$ implies $B_4^\wedge$. But

$$x \wedge (x' \vee 0) = (x \wedge x') \vee (x \wedge 0) = 0 \vee (x \wedge 0) = x \wedge 0 \ ,$$

$$x \wedge (x' \vee 0) = x \wedge x' = 0 \ ,$$

hence $x \wedge 0 = 0$, therefore

$$x \wedge (x \vee y) = (x \vee 0) \wedge (x \vee y) = x \vee (0 \wedge y) = x \vee 0 = x.$$

The original formulation of this self-dual system refers to an algebra $(B, \vee, \wedge)$ of type (2,2): instead of B_6 one postulates the existence of 0 and 1 satisfying B_6, while the last axiom requires B_{10} provided the elements 0 and 1 exist and are unique. Under this form, the independence of the system was proved by Huntington himself and reproved by Bernstein [1924] and later by Gerrish [1978], after a careful discussion, but apparently unaware of of Bernstein's proof. The same problem was tackled by Rüthing [1974]*; cf. MR 58(1979), #425. Sampathkumar [1967]* provided a version of system $\mathbf{B}_2$; cf. MR 50(1975), #12846. Diamond [1934a] realized the complete existential theory of the system.

It is Moore [1910] who defined the *complete existential theory* of a system of n conditions $p_1, \ldots, p_n$. By this term he meant the description of all the implications that hold between Boolean combinations of $p_1, \ldots, p_n$. In particular if there is no such implication, the system is said to be *completely independent*. Several complete existential theories occurring in the theory of partially ordered sets, set theory, lattice theory and the theory of Boolean algebras have been studied so far; see Appendix F.

Successive works of Del Re [1911]*, Bernstein [1914]* and Stone [1935b] yielded the independent system

$$\mathbf{B}_3 = \{B_1^\vee, B_2^\vee, B_2^\wedge, B_5^\wedge, B_6^\vee, B_6^\wedge, B_{10}\} \ ,$$

which is simpler than a previous system given by Del Re [1907].
Stone [1935b] devised the system

$$\mathbf{B}_4 = \{B_1^\vee, B_1^\wedge, B_2^\vee, B_5^\wedge, B_6^\vee, B_{11}^\wedge, B_{10}\}\,,$$

where

$B_{11}^\wedge \qquad (x \vee y) \wedge z = (x \wedge z) \vee (y \wedge z)\,.$

Newman's sets of axioms [1941] are essentially

$$\mathbf{B}_5 = \{B_1^\wedge, B_5^\wedge, B_{11}^\wedge, B_{12}^\vee, B_{10}, B_{13}^\vee\}\,,$$

$$\mathbf{B}_6 = \{B_1^\wedge, B_5^\wedge, B_{11}^\wedge, B_{12}^\wedge, B_{10}, B_{13}^\vee\}\,,$$

$$\mathbf{B}_7 = \{B_5^\wedge, B_{11}^\wedge, B_{12}^\vee, B_{12}^\wedge, B_{10}, B_{13}^\vee\}\,,$$

where

$B_{12}^\vee \qquad 0 \vee x = x\,,$

$B_{12}^\wedge \qquad 1 \wedge x = x\,,$

$B_{13}^\vee \qquad (z \vee z) \vee x = x \ \forall x \Longrightarrow z \vee z = x \ \forall x\,.$

In their original forms, systems $\mathbf{B}_5 - \mathbf{B}_7$ are independent.

The next three definitions are free from postulated special elements 0,1, meaning that not only the algebra is of type (2,2), but 0 and 1 do not occur in the axioms, their existence being proved from the axioms.

It is Bernstein [1915-16] who introduced the axiom

$B_{14} \qquad \forall x \exists x' \forall y \ \ y \vee (x \wedge x') = y \ \& \ y \wedge (x \vee x') = y$

and devised a self-dual 5-axiom system for Boolean lattices. As shown by Montague and J. Tarski [1954], one of the two axioms B_2 in that system is redundant and by deleting it one obtains an independent system, e.g.,

$$\mathbf{B}_8 = \{B_2^\vee, B_5^\vee, B_5^\wedge, B_{14}\}\,.$$

Diamond [1934b] provided the independent self-dual system

$$\mathbf{B}_9 = \{B_{15}^\vee, B_{15}^\wedge, B_{14}\}\,,$$

where

$B_{15}^\vee \qquad (y \wedge z) \vee x = (z \vee x) \wedge (y \vee x)$

and $B_{15}^\wedge$ is the dual axiom. Sholander obtained an even simpler system:

Theorem 4.1.1. (Sholander [1951]) *The system*

$$\mathbf{B}_{10} = \{\mathrm{B}_4^\wedge, \mathrm{B}_{16}^\wedge, \mathrm{B}_{17}\} \,,$$

where

$\mathrm{B}_{16}^\wedge$ $x \wedge (y \vee z) = (z \wedge x) \vee (y \wedge x)\,,$

B_{17} $\forall x \,\exists x' \,\forall y \;\; y \wedge (x \vee x') = y \vee (x \wedge x')$

characterizes Boolean algebras and is independent.

PROOF: First note that B is a distributive lattice by Theorem 3.2.1. Then B_{17} implies that $y \geq x \wedge x'$ and $x \vee x' \geq y$ for any y, showing that $x \wedge x' = 0$ (i.e., least element) and $x \vee x' = 1$ (i.e., greatest element).

To prove the independency consider e.g. by the following models. For $\mathrm{B}_4^\wedge$ take the set $\{0, 1\}$ endowed with the constant operations $x \vee y = x \wedge y = 1$. For $\mathrm{B}_{16}^\wedge$ take the non-modular lattice N5, change the notation to $o = 0$, $x = a$, $y = b$, $z = c$, $u = 1$ and define $0' = 1$, $a' = b' = c$, $c' = a$, $1' = 0$; then $x \vee x' = 1$ and $x \wedge x' = 0$ hold for all x, hence B_{17} holds. For B_{17} take any totally ordered set with more than two elements.   □

Starting from Huntington's first set, Van Albada [1964] derived 16 independent systems of axioms for Boolean lattices (within algebras $(B, \vee, \wedge)$), described by the following common wording:

A1. There is a one-sided identity 0 with respect to $\wedge$.

A2. There is a one-sided identity 1 with respect to $\vee$.

A3. $\wedge$ is one-sided distributive over $\vee$.

A4. $\vee$ is one-sided distributive over $\wedge$.

A5. A1 & A2 imply that one of the two identities is two-sided.

A6. A3 & A4 imply that one of the operations is two-sided distributive over the other.

A7. A1 & A2 imply that 0 and 1 can be chosen so that $\forall x \,\exists y \; x \vee y = 1$ & $x \wedge y = 0$.

It is understood that the choice is the same in A1 as in A2 and in A3 as in A4, so there are 16 instances of the above system A1-A7. Half of these specific systems are actually given in the paper with proofs, while the other half consists of their duals.

We conclude this section with the following natural problem: given a lattice which turns out to be the reduct of a Boolean algebra, is it true that that Boolean algebra is unique? The affirmative answer follows from the following more general theorem established by Wiener [1917]: the Boolean algebras $(B, \cup, \cap, {}^*, \alpha, \omega)$ defined on a Boolean algebra $(B, \vee, \wedge, {}', 0, 1)$ by Boolean (i.e., polynomial) operations $\cup, \cap, {}^*$ are of the form

$$x \cup y = (x \wedge y) \vee (a' \wedge (x \vee y)) \,,$$

$$x \cap y = (x \wedge y) \vee (a \wedge (x \vee y)) \,,$$

$$x^* = x' \,,$$

$$\alpha = 0, \ \omega = a' \,,$$

and they are isomorphic. See also Rudeanu [1974], Ch.12.

4.2. Boolean Algebras in Terms of the Operations ∨, ∧ and '

We begin this section with a kind of pattern for the passage from §1 to §2: an axiom system given by Bernstein [1950], which is the transform in terms of $\vee, \wedge, {}'$ of the earlier system $\mathbf{B}_8$ of the same author [1915/6] for Boolean lattices.

Then we give several sufficient conditions ensuring that a lattice (distributive lattice) endowed with a unary operation ', or ortholattice, or Newman algebra, is a Boolean algebra. Such results generate axiom systems for Boolean algebras from axiom systems for the larger class.

The last part of this section includes an independent self-dual system.

In the previous section we have seen that Bernstein [1915/6] devised a system of axioms in which 0,1 and complementation were not taken as basic operations. He then [1950] transformed the original system by working with algebras $(B, \vee, \wedge, {}')$ of type (2,2,1). So he replaced axiom B_{14} by

$B_{18}^{\vee}$ $x \vee (y \wedge y') = x \,,$

$B_{18}^{\wedge}$ $x \wedge (y \vee y') = x \,.$

Like for his earlier system, he believed that system $\{B_2^{\vee}, B_2^{\wedge}, B_5^{\vee}, B_5^{\wedge}, B_{18}^{\vee}, B_{18}^{\wedge}\}$ was independent, which is false, as shown by Montague and J. Tarski [1954]. After deletion of a redundant axiom, one obtains the independent system

$$\mathbf{B}_{11} = \{B_2^{\vee}, B_5^{\vee}, B_5^{\wedge}, B_{18}^{\vee}, B_{18}^{\wedge}\} \,.$$

More generally, Sioson [1964] found all the independent systems defining Boolean algebras that can be made out of the eight axioms

$B_2^\vee, B_2^\wedge, B_5^\vee, B_5^\wedge, B_{18}^\vee, B_{18}^\wedge,$

$B_{19}^\vee \qquad x \vee (y \vee y') = y \vee y'$,

$B_{19}^\wedge \qquad x \wedge (y \wedge y') = y \wedge y'$.

There are 10 such systems, grouped into pairs of dual systems; none of them is self-dual. Then Sioson [1967] extended this research by adding the axioms $B_1^\vee, B_1^\wedge, B_4^\vee, B_4^\wedge$.

Proposition 4.2.1. *A lattice endowed with a unary operation ' is a Boolean algebra if and only if it satisfies*

$B_{20} \qquad (x \wedge y) \vee (x \wedge y') = (x \vee y) \wedge (x \vee y')$.

PROOF: Since in every Boolean algebra

$B_{21}^\vee \qquad (x \wedge y) \vee (x \wedge y') = x$,

$B_{21}^\wedge \qquad (x \vee y) \wedge (x \vee y') = x$,

it remains to prove sufficiency. But B_{20} implies B_{21} because in any lattice

$$(x \wedge y) \vee (x \wedge y') \leq x \leq (x \vee y) \wedge (x \vee y') .$$

Now take x, y, z such that $x \vee z = y \vee z$ and $x \wedge z = y \wedge z$. Then

$$x = (x \wedge z) \vee (x \wedge z') = (y \wedge z) \vee ((x \vee z) \wedge (x \vee z') \wedge z') = (y \wedge z) \vee ((x \vee z) \wedge z')$$

$$= (y \wedge z) \vee ((y \vee z) \wedge z' \wedge (y \vee z')) = (y \wedge z) \vee (y \wedge z') = y ,$$

therefore the lattice is distributive by Corollary 3.1.3. This implies

$$x \wedge (y \vee y') = (x \wedge y) \vee (x \wedge y') = x ,$$

showing that $y \vee y' = 1$ is greatest element and similarly $x \wedge x' = 0$ is least element. □

Corollary 4.2.1. *Boolean algebras are characterized by the independent self-dual set of three identities $L_{62}^\vee$, $L_{62}^\wedge$ and B_{20}.*

PROOF: The system characterizes Boolean algebras by the above Proposition and Theorem 1.2.5.

The independence of axiom B_{20} is shown by any lattice with 0 in which $y' = 0$ for all y. Now, in view of duality, it suffices to prove the independence of $L_{62}^\vee$.

Consider the set $\{0, 1, 2\}$ endowed with the operations $\wedge, \vee, '$ given by

$$x \wedge 0 = 0, \ x \wedge 1 = x, \ x \wedge 2 = 2 ,$$

$$0 \vee 0 = 0, \ x \vee 1 = 1 \vee x = x, \ 0 \vee 2 = 2 \vee 0 = 1, \ 2 \vee 2 = 2 ,$$

$$0' = 1, \ 1' = 2, \ 2' = 1 \ .$$

Axiom $L_{26}^{\vee}$ fails for $y := 0, u := 1, w := 2$, for which $1 \wedge ((v \wedge 0) \vee (0 \wedge 2)) = 1$, hence the left-hand side of $L_{26}^{\vee}$ is 1.

Taking in turn $y := 0, 1, 2$, axiom B_{20} reduces to $x \vee 0 = 0 \vee x, x \vee 2 = x \vee 2$ (because $1 \wedge a = a$) and $x \vee 2 = 2 \vee x$, respectively.

It remains to check $L_{26}^{\wedge}$. Taking in turn $y := 0, 1, 2$, we obtain $((x \vee y) \wedge y) \vee (z \wedge y) = y$, therefore $L_{26}^{\wedge}$ reduces to

L_{26}' $\qquad y \wedge (u \vee t) = y$, where $t = (v \vee y) \wedge (y \vee w)$.

L_{26}' is readily checked for $y := 1$.

Further note that $v \vee 0, 0 \vee w \in \{0, 1\}$ and $\{0, 1\}$ is a subalgebra of our structure, hence for $y := 0$ we have $t \in \{0, 1\}$, therefore $u \vee t \in \{0, 1\}$ (easy remark), hence $0 \wedge (u \vee t) = 0$.

Similarly, $v \vee 2, 2 \vee w \in \{1, 2\}$ and $\{1, 2\}$ is a subalgebra, hence for $y := 2$ we have $t \in \{1, 2\}$, therefore $u \vee t \in \{1, 2\}$, hence $2 \wedge (u \vee t) = 2$. $\square$

Proposition 4.2.2. *A lattice L endowed with a unary operation $'$ is a Boolean algebra if and only if it satisfies*

$B_{22}^{\vee}$ $\qquad ((x \vee y') \wedge y) \vee (x \wedge y') = x$,

$B_{22}^{\wedge}$ $\qquad ((x \wedge y') \vee y) \wedge (x \vee y') = x$.

PROOF: Taking in $B_{22}^{\wedge}$ the meet of each side with y, we get $(x \vee y') \wedge y = x \wedge y$, therefore $B_{22}^{\vee}$ reduces to $B_{21}^{\vee}$. We obtain similarly $B_{21}^{\wedge}$, therefore B_{20} holds and we apply the previous proposition. $\square$

Propositions 2.3 and 2.4 below characterize Boolean algebras among the class of all lattices of type (2,2,1) using the idea of the uniqueness of Mal'cev terms (see D. Kelly and R. Padmanabhan [2007] for more details).

Proposition 4.2.3. *A lattice endowed with a unary operation $'$ is Boolean if and only if it satisfies the self-dual identity*

$B_1^{\vee\wedge}$ $\qquad (x \wedge y \wedge z) \vee (x \wedge y' \wedge z') \vee (x' \wedge y \wedge z') \vee (x' \wedge y' \wedge z)$

$$= (x \vee y \vee z) \wedge (x \vee y' \vee z') \wedge (x' \vee y \vee z') \wedge (x' \vee y' \vee z) \ .$$

PROOF: In a Boolean algebra the left-hand side of $B_1^{\vee\wedge}$ is clearly $x + y + z$, while the right-hand side equals

$$(x \vee (y \wedge z') \vee (y' \wedge z)) \wedge (x' \vee (y \wedge z) \vee (y' \wedge z')) = (x \vee (y + z)) \wedge (x' \vee (y + z)')$$

$$= (x \wedge (y + z)') \vee (x' \wedge (y + z)) = x + y + z \ .$$

Conversely, suppose identity $B_1^{\vee\wedge}$ holds. Taking $z := y$ and using well-known lattice identities, we obtain

$$x \geq (x \wedge y) \vee (x \wedge y') = (x \vee y)\wedge)x \vee y') \geq x ,$$

hence identities

$$(x \wedge y) \vee (x \wedge y') = x ,$$

$$(x \vee y) \wedge (x \vee y') = x ,$$

hold, therefore the lattice is a Boolean algebra by Proposition 2.1.   □

Proposition 4.2.4. *A lattice endowed with a unary operation ' is a Boolean algebra if and only if it satisfies the self-dual identity*

$B_2^{\vee\wedge}$ $(x \vee (x \vee z)') \wedge (y \vee (y \vee z)') \wedge (x \vee z)$
$$= (x \wedge (x \wedge z)') \vee (y \wedge (y \wedge z)') \vee (x \wedge y) .$$

PROOF: In a Boolean algebra we have

$$x \vee (x \vee z)' = x \vee (x' \wedge z') = x \vee z'$$

and dually, therefore identity $B_2^{\vee\wedge}$ follows by Corollary 3.1.1.

Conversely, suppose identity $B_1^{\vee\wedge}$ holds. Taking $z := x$ and using well-known lattice identities, we obtain

$$y \leq (y \vee (y \vee x)') \wedge (x \vee y) = (y \wedge (y \wedge x)') \vee (x \wedge y) \leq y ,$$

therefore

(1) $(y \vee (y \vee x)') \wedge (x \vee y) = x ,$

(2) $(y \wedge (y \wedge x)') \vee (x \wedge y) = y .$

Besides, taking $x := 1$, $z := x$ in $B_2^{\vee\wedge}$ we get

$$y \vee (y \vee x)' = x' \vee (y \wedge (y \wedge x)') \vee y = x' \vee y$$

and $y \wedge (y \wedge x)' = x' \wedge y$, hence (1) and (2) reduce to $B_{21}^{\wedge}$ and $B_{21}^{\vee}$ (with x and y interchanged), which imply B_{20}, therefore the lattice is a Boolean algebra by Proposition 2.1.   □

Remark 4.2.1. A distributive lattice is a Boolean algebra if and only if it satisfies

$B_{23}^{\vee}$ $x \vee x' = y \vee y' ,$
$B_{23}^{\wedge}$ $x \wedge x' = y \wedge y' ,$

because $x \vee x'$ and $x \wedge x'$ are constants and in view of Remark 4.1.1 we can safely denote them by $x \vee x' = 1$ and $x \wedge x' = 0$.

An ortholattice is a complemented bounded lattice (not necessarily distributive) satisfying the De Morgan laws and the law of double negation. An ortholattice is a Boolean algebra if and only if it satisfies $B_{21}^{\vee}$. This led to the following characterization of Boolean algebras:

Theorem 4.2.1. (Beran [1982]) *The following system characterizes Boolean algebras:*

$$\mathbf{B}_{12} = \{B_{18}^{\vee}, B_{24}^{\vee}, B_{25}^{\vee}\},$$

where

$B_{24}^{\vee}$ $(x \vee y) \vee z = (z' \wedge y')' \vee x$,

$B_{25}^{\vee}$ $(x \wedge (y \vee z)) \vee (x \wedge y') = x$.

PROOF: Clearly the axioms $\mathbf{B}_{12}$ are fulfilled in a Boolean algebra. Conversely, note that $(x \wedge (x \vee z)) \vee (x \wedge x') = x \wedge (x \vee z)$ by $B_{18}^{\vee}$ and $(x \wedge (x \vee z)) \vee (x \wedge x') = x$ by $B_{25}^{\vee}$, therefore $x \wedge (x \vee z) = x$. But this identity together with $B_{18}^{\vee}$ and $B_{24}^{\vee}$ characterize ortholattices; cf. Beran [1976]. On the other hand, taking $z := y \wedge y'$ in $B_{25}^{\vee}$ and using again $B_{18}^{\vee}$, we obtain the identity $B_{21}^{\vee}$, which characterizes Boolean algebras within ortholattices.
□

The following generalization of Boolean algebras was introduced by Newman [1941]: an algebra $(A, \vee, \wedge, 0, 1)$ of type $(2,2,0,0)$ satisfying the axioms $B_5^{\wedge}, B_6^{\vee}, B_6^{\wedge}, B_{10}$ and $0 \vee x = x$; the element x' in B_{10} is in fact uniquely determined by x. This structure is known as a *Newman algebra* and it is isomorphic to the Cartesian product of the subalgebra of the elements satisfying $x \vee x = x$ and the subalgebra of the elements satisfying $x \vee x = 0$; the former subalgebra is a Boolean algebra. So Boolean algebras coincide with Newman algebras satisfying the identity $x \vee x = x$.

There are many axiom systems for Newman algebras; see e.g. G.D. Birkhoff and G. Birkhoff [1946], Wooyenaka [1964], Sioson [1965a], [1965b], [1967], Sobociński [1972c], [1972d], [1972e], [1973]. Sioson [1967] has proved that each of the following identities characterizes Boolean algebras among Newman algebras: $B_1^{\vee}, B_4^{\vee}, B_4^{\wedge}, B_5^{\vee}, B_{19}^{\vee}$. As a matter of principle, this generates even more systems of postulates for Boolean algebras; some of them have been actually written down. We refer to these systems here because they characterize Boolean algebras within a larger class, but note that from the point of view of axiom B_{10} they should have been presented in §1. The first system obtained under the influence of Newman algebras is due to Birkhoff [1948]:

$$\mathbf{B}_{13} = \{B_1^\vee, B_5^\vee, B_5^\wedge, B_6^\vee, B_{12}^\vee B_{12}^\wedge, B_8^\vee, B_8^\wedge\}\ .$$

A quite different idea for generating a system of axioms was suggested by G.D. Birkhoff and G. Birkhoff [1946]. They noted that the axioms obtained from $B_5^\wedge, B_6^\wedge, B_8^\vee$ by applying the transformations of the 8-element group generated by left-right symmetry and duality form a (redundant) system defining Boolean algebras. This is related to their philosophic view that "the final form of any scientific theory T is (1) based on a few simple postulates, and (2) contains an extensive ambiguity, associated symmetry, and underlying group G, in such wise that ... T appears nearly self-evident in view of the Principle of Sufficient Reason".

The problem of finding an independent self-dual system of identities for Boolean algebras was raised by Grätzer [1971], Problem 29 (Huntington's first system and Diamond's system $\mathbf{B}_9$ are not equational). The first answer to this problem was given in

Theorem 4.2.2. (Padmanabhan [1983]) *The self-dual system*

$$\mathbf{B}_{14} = \{B_{26}^\vee, B_{26}^\wedge, B_{27}^\vee, B_{27}^\wedge, B_8^\vee, B_8^\wedge\}\ ,$$

where

$B_{26}^\vee$ $\quad (x \wedge y) \vee y = y\ ,$

$B_{26}^\wedge$ $\quad (x \vee y) \wedge y = y\ ,$

$B_{27}^\vee$ $\quad x \vee (y \wedge z) = (y \vee x) \wedge (z \vee x)\ ,$

$B_{27}^\wedge$ $\quad x \wedge (y \vee z) = (y \wedge x) \vee (z \wedge x)\ ,$

characterizes Boolean algebras and is independent.

PROOF: Clearly the well-known proof that the absorption laws imply the idempotency laws (see e.g. Ch.1,§1, proof for $\mathbf{L}_1$) works also for the form B_{26} of absorption. Now $B_1^\wedge, B_{27}^\vee$ and $B_1^\wedge$ imply

$$x \vee y = x \vee (y \wedge y) = (y \vee x) \wedge (y \vee x) = y \vee x\ ,$$

and similarly $x \wedge y = y \wedge x$. Therefore axiom $B_{26}^\vee$ coincides with $B_4^\vee$, while the distributivity $B_{27}^\wedge$ can be written in the form $x \wedge (y \vee z) = (z \wedge x) \vee (y \wedge x)$. This identity and $B_4^\wedge$ ensure that the algebra is a distributive lattice by

Sholander's Theorem 3.2.1, therefore it is a Boolean algebra by Remark 1.1.

The independence is shown by the following models. For $B_{26}^{\vee}$ take the set $\{0, 1\}$ with the operations $x \vee y = 1$, $x \wedge y = y$, $x' = 0$. For $B_8^{\vee}$ take the chain $\{0, 1\}$ with its lattice operations and define $x' = 0$. For $B_{27}^{\vee}$ take a chain $\{\alpha, \omega\}$, where $\alpha < \omega$, and define $x \vee y = y$, $x \wedge y = \min(x, y)$, $x' = \alpha$, $0 = 1 = \alpha$. For the dual axioms take the dual models. □

The fact that in lattice theory each distributive law implies the other makes it hard to produce an independent basis containing both distributive laws. Diamond's self-dual independent system $\mathbf{B}_9$ realizes this by axioms $B_{15}^{\vee}$ and $B_{15}^{\wedge}$, but uses axiom B_{14}, which is not an identity. If we replace B_{14} by the corresponding identities

$B_{14}^{\vee}$ $x \vee (y \wedge y') = x$,

$B_{14}^{\wedge}$ $x \wedge (y \vee y') = x$,

we obtain the following variant of Diamond's system:

Theorem 4.2.3. *The set* $\{B_{14}^{\vee}, B_{14}^{\wedge}, B_{15}^{\vee}, B_{15}^{\wedge}\}$ *is an independent self-dual basis for Boolean algebras.*

PROOF: Suppose an algebra $(L, \vee, \wedge, ')$ satisfies the above system. By using in turn $B_{14}^{\vee}, B_{15}^{\wedge}$ and again $B_{14}^{\vee}$, we obtain

$$x \wedge y = (x \wedge y) \vee (u \wedge u') = (y \vee (u \wedge u')) \wedge (x \vee (u \wedge u')) = y \wedge x$$

and similarly $x \vee y = y \vee x$, so that in the following we can freely use commutativity.

It follows by $B_{14}^{\vee}, B_{15}^{\vee}$ and $B_{14}^{\vee}$ that

(3)
$$\begin{aligned} x \wedge (y \wedge y') &= (x \wedge (y \wedge y')) \vee (x \wedge x') = x \wedge ((y \wedge y') \wedge y) \\ &= x \vee (y \wedge y') = x. \end{aligned}$$

Now it follows by $B_{14}^{\vee}, B_{15}^{\wedge}, (3)$ and $B_{14}^{\vee}$, that

$$x \wedge (x \vee y) = (x \vee (y \wedge y')) \wedge (x \vee y) = x \vee ((y \wedge y') \wedge y) = x \vee (y \wedge y') = x$$

and similarly $x \vee (x \wedge y) = x$. Therefore the reduct $(L, \vee, \wedge)$ is a distributive lattice by Sholander's Theorem 3.2.1. Now axioms $B_{14}^{\vee}$ and $B_{14}^{\wedge}$ read $y \wedge y' \leq x \leq y \vee y'$, showing that $y \wedge y'$ and $y \vee y'$ are the least and greatest elements of the lattice L, hence they are constants and by setting $y \wedge y' = 0$ and $y \vee y' = 1$ we get a Boolean algebra $(L, \vee, \wedge, ', 0, 1)$.

To prove the independence of the system, let $(B, \vee, \wedge, ', 0, 1)$ be an arbitrary Boolean algebra. Set $x - y = x \wedge y'$ and $x \Delta y = (x \wedge y) \vee (x' \wedge y')$.

Consider the algebra $(B, 0, -, ')$ of type $(2,2,1)$ (so $x \vee y = 0$ for all x, y). This algebra does not fulfil $B_{14}^{\vee}$ because $0 \neq 1$, while the other axioms aresatisfied because $x - 0 = x$, $0 = 0 - 0$ and $0 - x = 0$.

The algebra $(B, \vee, \Delta, ')$ of type $(2,2,1)$ does not fulfil $B_{15}^{\vee}$ because $(x \vee y)\Delta 0 = x' \wedge y'$ but $(y\Delta 0) \vee (x\Delta 0) = x' \vee y'$, while the other axioms are satisfied because

$$x \vee (y\Delta y') = x \vee 0 = x ,$$

$$x\Delta(y \vee y') = x\Delta 1 = x ,$$

$$(x\Delta y) \vee z = (x \wedge y) \vee (x' \wedge y') \vee z ,$$

$$(y \vee z)\Delta(x \vee z) = ((y \vee z) \wedge (x \vee z)) \vee ((y' \wedge z) \wedge (x' \wedge z'))$$

$$= (x \wedge y) \vee z \vee (x' \wedge y' \wedge z') = (x \wedge y) \vee z \vee (x' \wedge y') .$$

The duals of the above models establish the independence of the axioms $B_{14}^{\wedge}$ and $B_{15}^{\wedge}$, respectively.   □

We conclude with a few systems of axioms that are expressed in terms of $\vee, \wedge, '$ but do not fall under the general ideas emphasized above.

Ponticopoulos [1962] suggested a system consisting of $B_2^{\vee}, B_2^{\wedge}, B_9$ and an additional very long axiom.

Sampathkumar [1963] provided the system

$$\mathbf{B}_{15} = \{B_2^{\vee}, B_2^{\wedge}, B_5^{\wedge}, B_6^{\wedge}, B_7^{\vee}, B_8^{\vee}, B_8^{\wedge}\} ,$$

Carloman [1976] proposed the system

$$\mathbf{B}_{16} = \{B_2^{\vee}, B_4^{\vee}, B_5^{\vee}, B_8^{\vee}\} ,$$

while Lisovik [1977] devised the system

$$\mathbf{B}_{17} = \{B_2^{\vee}, B_5^{\vee}, B_{28}^{\wedge}, B_{29}^{\vee}\} ,$$

where

$B_{28}^{\wedge}$ $x \wedge y = (x' \vee y')' ,$

$B_{29}^{\vee}$ $(y \wedge y') \vee x = x .$

4.3. Boolean Algebras in Terms of the Operations $\vee$ and $'$

The first definition of Boolean algebras in terms of join and complements is due to Huntington [1904] and is known as *Huntington's third set*. In modern terms, he considered a bounded join semilattice satisfying two more postulates: 1) for each element x there is an element x' such that $x \vee x' = 1$ and the unique lower bound of x and x' is 0, and 2) every two non-0 elements x and y have a lower bound $z \neq 0$.

The other systems of axioms presented in this section refer to algebras B endowed with a binary operation $\vee$ and possibly with a constant 0 or a constant 1 or both. Some of these systems include a subset equivalent to the fact that $(B, \vee)$ is a semilattice, others do not; in all cases the operation $\vee$ is taken to be the join semilattice of the Boolean algebra. If the type of the original algebra includes only one of the constants 0,1, the other constant is defined by complementation. If the type of the algebra is just $(2,1)$, the elements 0 and 1 are obtained as in Remark 2.1 or alike. In all cases the meet operation is defined by [†]

$$(1) \qquad\qquad x \wedge y = (x' \vee y')' \ ;$$

the point is that in a Boolean algebra identity (1) holds by (1.2) and B_9.

The first equational definitions of Boolean algebras are again due to Huntington [1933a] and known as his *fourth* and *fifth* systems, namely the independent systems

$$\mathbf{B}_{18} = \mathbf{IV} = \{B_2^\vee, B_3^\vee, B_{30}^\vee\} \ ,$$

where

$B_{30}^\vee \qquad (x' \vee y')' \vee (x' \vee y)' = x \ ,$

and

$$\mathbf{B}_{19} = \mathbf{V} = \{B_9, B_{31}^\vee, B_{32}^\vee\} \ ,$$

suggested by Sheffer's system $\mathbf{B}_{52}$, where

$B_{31}^\vee \qquad x \vee (y \vee y')' = x \ ,$

$B_{32}^\vee \qquad ((y' \vee x)' \vee (z' \vee x)')' = x \vee (y \vee z)' \ .$

Of course, the commutativity $B_2^\vee$ is obtained due to the "reverse" order in the left-hand side of $B_{32}^\vee$. However we can prove commutativity even without $B_{31}^\vee$ and with the "normal order" in $B_{32}^\vee$, provided two other simple axioms are introduced. This is shown by the system

[†]Note that (1) is a definition, not to be confused with Lisovik's axiom $B_{28}^\wedge$.

$\mathbf{B}'_{20} = \{B_6^\vee, B_9, B_{33}^\vee, B_{34}^\vee\}$, where

$B_{33}^\vee \qquad x \vee x' = 0'$,

$B_{34}^\vee \qquad ((x \vee y') \vee (x \vee z')')' = x \vee (y \vee z)'$.

Rudeanu [1963], following a suggestion of Moisil, proved that system

$$\mathbf{B}_{20} = \mathbf{B}'_{20} \cup \{B_2^\vee\}$$

characterizes Boolean algebras, as well as the independence of the axioms in $\mathbf{B}_{20}$, and asked whether $B_2^\vee$ can be deduced from $\mathbf{B}_{20}$. Petcu [1967] answered in the affirmative.

On the other hand, the commutativity of the join can replace the law of double negation in Huntington's fifth set:

Proposition 4.3.1. (Lisovik [1997]) *The system*

$$\mathbf{B}_{21} = \{B_2^\vee, B_{31}^\vee, B_{32}^\vee\}$$

characterizes Boolean algebras and is independent.

PROOF: It suffices to derive B_9 from $\mathbf{B}_{21}$. First we note that

$$((x'' \vee x)' \vee (x''' \vee x)')' = x \vee (x' \vee x'')' = x ,$$

showing that every element x can be written in the form $x = w'$. Now set $(x \vee x')' = z$. Using freely the commutativity, we compute

$$x = z \vee x = z \vee w' = z \vee (z \vee w)'$$

$$= ((z' \vee z)' \vee (w' \vee z)')' = ((w' \vee z)')' = w''' = x'' .$$

The independence is left to the reader.   □

The fact that Boolean algebra is the algebraic counterpart of classical propositional calculus raises the possibility of transforming systems of axioms for the latter theory into alternative definitions of Boolean algebras. For instance, Bernstein [1931] transcribed a well-known formal system of propositional calculus into the language of an algebra $(B, \vee, ', 1)$, e.g.

$$p = 1 \ \& \ p' \vee q = 1 \Longrightarrow q = 1$$

was the transcription of modus ponens, while the axiom $\vdash p \to (p \vee q)$ became $p' \vee (p \vee q) = 1$; etc. Henle [1932] noted that the system obtained in this way did not suffice to characterize Boolean algebras. Then Huntington [1933a] corrected the error and obtained an independent system known as his *sixth* set. Bennett [1933] simplified one of the axioms and obtained a new independent system.

Another system originating in propositional calculus was given by Lewis and Langford [1932], namely

$$\mathbf{B}_{22} = \{\mathrm{B}_1^\vee, \mathrm{B}_2^\vee, \mathrm{B}_3^\vee, \mathrm{B}_7^\vee, \mathrm{B}_{35}^\vee, \mathrm{B}_{36}^\vee\}\,,$$

which defines Boolean algebras as join semilattices with greatest element 1 and a unary operation ′ satisfying

$\mathrm{B}_{35}^\vee$ $x \vee y' = 1 \Longrightarrow x \vee y = x\,,$

$\mathrm{B}_{36}^\vee$ $x \vee y = x\ \&\ x \vee y' = x \Longrightarrow x = 1\,.$

However Rudeanu [1962] observed that axiom $\mathrm{B}_7^\vee$ is redundant (because taking $x := 1 \vee 1'$, $y := 1$ in $\mathrm{B}_{36}^\vee$ yields $1 \vee 1' = 1$, then the same axiom with $x := x \vee 1$, $y = 1$ implies $x \vee 1 = 1$), while the system

$$\mathbf{B}_{23} = \{\mathrm{B}_1^\vee, \mathrm{B}_2^\vee, \mathrm{B}_3^\vee, \mathrm{B}_{35}^\vee, \mathrm{B}_{36}^\vee\}$$

is independent.

In Jarbuch Fortschritte Math. 59(1933), p.59, it is stated that Huntington [1933b]* improved an axiom system given by Lewis and Langford.

The following important theorem is due to Frink [1941], with a proof which invokes the axiom of choice and Zorn's lemma. An elementary proof was given by Padmanabhan [1981], which is essentially the one given below.

Theorem 4.3.1. *An algebra* $(B, \vee, ')$ *can be made into a Boolean algebra if and only if* $(B, \vee)$ *is a semilattice which satisfies*

$\mathrm{B}_{37}^\vee$ $x \vee y = y \Longleftrightarrow x' \vee y = z' \vee z\,.$

PROOF: Since $x' \vee x = z' \vee z$ by $\mathrm{B}_{37}^\vee$, it follows by Remarks 2.1 and 1.1 that $1 = x \vee x'$ is greatest element, hence $\mathrm{B}_{37}^\vee$ reads

$$(2) \qquad\qquad x \vee y = y \Longleftrightarrow x' \vee y = 1\,.$$

Taking $y := x''$, this implies further

$$(3) \qquad\qquad x \vee x'' = x''$$

and in particular $x' \vee x''' = x'''$, hence

$$x''' \vee x = x \vee x''' = x \vee x' \vee x''' = 1 \vee x''' = 1\,,$$

therefore (2) implies $x'' \vee x = x$; comparing to (3), we obtain B_9.

Define $0 = 1'$ and $x \wedge y = (x' \vee y')'$. This immediately implies, via B_9, the De Morgan laws, whence it is readily checked that $\wedge$ is a semilattice operation as well, and using also (2) we get

$$x \wedge y = x \iff (x \wedge y)' = x' \iff x' \vee y' = x' \iff y' \vee x' = x'$$

$$\iff y'' \vee x' = 1 \iff x' \vee y = 1 \iff x \vee y = y \,,$$

therefore $(B, \vee, \wedge)$ is a lattice with order relation

(4) $x \leq y \iff x \wedge y = x \iff x \vee y = y \iff x' \vee y = 1 \iff x \wedge y' = 0.$

Besides, $x \vee x' = 1$, hence $x \wedge x' = x' \wedge x = 1' = 0$, therefore B is a complemented lattice by Remark 1.1.

To prove distributivity we note that

$$x \wedge (x \wedge y)' \wedge y'' = x \wedge y \wedge (x \wedge y)' = 0 \,,$$

whence (4) implies $x \wedge (x \wedge y)' \leq y'$. We obtain similarly $x \wedge (x \wedge z)' \leq z'$, therefore

$$x \wedge (x \wedge y)' \wedge (x \wedge z)' \leq y' \wedge z' = (y \vee z)'$$

and taking the meet of each side with $y \vee z$ we obtain

$$x \wedge (y \vee z) \wedge ((x \wedge y)' \wedge (x \wedge z)')'' = 0 \,,$$

hence

$$x \wedge (y \vee z) \leq ((x \wedge y)' \wedge (x \wedge z)')' = (x \wedge y) \vee (x \wedge z)$$

and since the converse inequaliy holds in any lattice, it follows that $x \wedge (y \vee z) = (x \wedge y) \vee (x \wedge z)$. $\square$

This theorem has many consequences. Note first that since universal quantifiers are understood in $B_{37}^{\vee}$, this axiom says in fact the following:
        either $x \vee y = y$ is true and $x' \vee y = z' \vee z$ is true for all z,
        or $x \vee y = y$ is false and $x' \vee y = z' \vee z$ is false for all z.
Now the following pair of axioms has the same meaning:

$B_{38}^{\vee}$ $x \vee y = y \implies \forall z \, (x' \vee y = z' \vee z) \,,$

$B_{39}^{\vee}$ $\exists z \, (x' \vee y = z' \vee z) \implies x \vee y = y \,.$

Byrne [1946] proved that the idempotency of the operation $\vee$ can be dropped in Frink's theorem, while Goodstein [1963] re-proved this result[†], working with axioms $B_{38}^{\vee}, B_{39}^{\vee}$; he provided the dual of the system

$$\mathbf{B}_{24} = \{B_2^{\vee}, B_3^{\vee}, B_{38}^{\vee}, B_{39}^{\vee}\} \,.$$

[†]With no reference to Frink or Byrne.

Rudeanu [1962], analyzing Byrne's proofs, observed that in fact they constructed the duals of the systems $\mathbf{B}_{24}$ and

$$\mathbf{B}_{25} = \{\mathrm{B}_{38}^{\vee}, \mathrm{B}_{39}^{\vee}, \mathrm{B}_{40}^{\vee}\}\,,$$

where

$\mathrm{B}_{40}^{\vee}$ $(x \vee y) \vee z = (y \vee z) \vee x$.

We give below some other corollaries of Frink's theorem. Their proofs have a common scheme. First one shows that B is a semilattice having as greatest element the constant $z \vee z' = 1$. Then one checks $\mathrm{B}_{37}^{\vee}$ in the form (2).

Theorem 4.3.2. *An algebra $(B, \vee, \,')$ can be made into a Boolean algebra if and only if $(B, \vee)$ is a semilattice which satisfies* $\mathrm{B}_9, \mathrm{B}_{23}^{\vee}$ *and*

$\mathrm{B}_{41}^{\vee}$ $(x' \vee y)' \vee (x \vee y)' = y'$.

PROOF: We set $x \vee x' = 1$, which is a constant by $\mathrm{B}_{23}^{\vee}$. Then taking $y := x'$ in $\mathrm{B}_{41}^{\vee}$, seting $1' = 0$ and using B_9 we obtain $x \vee 0 = x$. Besides, $x \vee 1 = x \vee x \vee x' = x \vee x' = 1$. Now (2) follows easily via $\mathrm{B}_{41}^{\vee}$ and B_9:

$$x' \vee y = 1 \Longrightarrow 0 \vee (x \vee y)' = y' \Longrightarrow x \vee y = y\,,$$

$$x \vee y = y \Longrightarrow x' \vee y = x' \vee x \vee y = 1 \vee y = 1\,.$$

$\square$

Proposition 4.3.2. (Malliah [1968]) *The following system characterizes Boolean algebras within algebras $(B, \vee, \,', 0)$:*

$$\mathbf{B}_{26} = \{\mathrm{B}_9, \mathrm{B}_{33}^{\vee}, \mathrm{B}_{42}^{\vee}, \mathrm{B}_{43}^{\vee}\}\,,$$

where

$\mathrm{B}_{42}^{\vee}$ $x \vee (y \vee z) = z \vee (x \vee y)$,

$\mathrm{B}_{43}^{\vee}$ $x \vee y = 0' \Longrightarrow x \vee y' = x$.

PROOF: It follows from $\mathrm{B}_{43}^{\vee}$ that $x \vee x' = 0' \Longrightarrow x \vee x'' = x$, whence $\mathrm{B}_{33}^{\vee}$ and B_9 imply $x \vee x = x$. But this property and $\mathrm{B}_{42}^{\vee}$ are S_1 and S_7 in Ch.1 §1, therefore $(B, \vee)$ is a semilattice and setting $0' = 1$, axiom $\mathrm{B}_{33}^{\vee}$ reads $x \vee x' = 1$. This implies that $x \leq 1$, that is, 1 is greatest element. Now we check (2):

$$x \vee y = y \Longrightarrow x' \vee y = x' \vee x \vee y = 1 \vee y = 1\,,$$

while from $B_{43}^\vee$ and B_9 we get

$$x' \vee y = 1 \implies y \vee x' = 0' \implies y \vee x'' = y \implies x \vee y = y \;.$$

$\square$

Proposition 4.3.3. *The following system characterizes Boolean algebras within algebras* $(B, \vee, ')$:

$$\mathbf{B}_{27} = \{B_1^\vee, B_{44}^\vee\} \;,$$

where

$B_{44}^\vee \qquad ((x \vee y) \vee z) \vee t = (y \vee z) \vee x \iff ((x \vee y) \vee z) \vee t' = u \vee u' \;.$

PROOF: It follows from $B_1^\vee$ that

$$((x \vee x) \vee x) \vee x = x = (x \vee x) \vee x \;,$$

therefore $B_{44}^\vee$ implies $x \vee x' = u \vee u'$, showing that $x \vee x' = 1$ is a constant. Now taking $t := (x \vee y) \vee z$ in $B_{44}^\vee$ and using $B_1^\vee$, we get $(x \vee y) \vee z = (y \vee z) \vee x$. But this property and $B_1^\vee$ are S_6 and S_1 in Ch.1 §1, therefore $(B, \vee)$ is a semilattice. Taking $y := z := x$ in $B_{44}^\vee$, we obtain $x \vee t = x \iff x \vee t' = 1$, which coincides with (2) due to commutativity. $\square$

A two-axiom equational characterization of Boolean algebras was provided by Sobociński [1979]:

Proposition 4.3.4. *The following system characterizes Boolean algebras within algebras* $(B, \vee, ')$:

$$\mathbf{B}_{28} = \{B_1^\vee, B_{45}^\vee\} \;,$$

where

$B_{45}^\vee \qquad ((z' \vee t)' \vee (z' \vee t')') \vee (x \vee y) = x \vee (y \vee z) \;.$

PROOF: Since

$$x = x \vee x = ((x' \vee t)' \vee (x' \vee t')') \vee (x \vee x) \;,$$

it follows that

(5) $$x = w \vee x \;,$$

where we have set $(x' \vee t)' \vee (x' \vee t')' = w$. Now by applying in turn (5), $B_{45}^\vee$ with $x := y := w$, $z := x$, we obtain

$$x = w \vee (w \vee x) = ((x' \vee t)' \vee (x' \vee t')') \vee (w \vee w) = w \vee (w \vee w) = w \;,$$

that is,

(6) $$x = (x' \lor t)' \lor (x' \lor t')' \,,$$

which transforms $B_{45}^{\lor}$ into $z \lor (x \lor y) = x \lor (y \lor z)$. So the axioms system $\{S_1, S_7\}$ of semilattices is fulfilled, therefore the commutativity reduces (6) to $B_{30}^{\lor}$ and Huntington's fourth system $\mathbf{B}_{18}$ is fulfilled. $\quad\square$

Another two-identity characterization of Boolean algebras was given by Meredith and Prior [1968], namely

$$\mathbf{B}_{29} = \{B_{46}^{\lor}, B_{47}^{\lor}\} \,,$$

where

$B_{46}^{\lor}$ $(x' \lor y)' \lor x = x \,,$

$B_{47}^{\lor}$ $(x' \lor y)' \lor (z \lor y) = y \lor (z \lor x) \,.$

In the early 1930s Robbins conjectured that in Huntington's fourth set $\mathbf{B}_{18}$ one can replace axiom $B_{30}^{\lor}$ by the following variant:

$B_{48}^{\lor}$ $((x \lor y)' \lor (x \lor y')')' = x \,,$

which became known as *Robbins' axiom*. So his conjecture was that system

$$\mathbf{B}_{30} = \{B_2^{\lor}, B_3^{\lor}, B_{48}^{\lor}\}$$

characterizes Boolean algebras. Winker [1992] proved that this would be true provided $\mathbf{B}_{29}$ implies the solvability of the equation $(x \lor y)' = y'$. McCune [1997], using the theorem prover Otter, solved the problem in the affirmative, showing that Winker's equation has solutions. Otter's proof contains fairly complex terms which are hard to understand or even to print in a readable format. Dahn [1998] obtained a quite readable "anthropomorphized" version of the proof. Other attempts in this direction are referred to in Dahn's paper.

The program Otter was also used by Phillips and Vojtěchovský [2005]*, who gave the system

$$\mathbf{B}_{31} = \{B_{48}^{\lor}, B_{49}^{\lor}\} \,,$$

where

$B_{49}^{\lor}$ $(x \lor y) \lor z = y \lor (z \lor x) \,,$

and by McCune, Veroff, Fittelson, Harris, Feist and Wos [2002], who found ten single-identity characterizations of Boolean algebras. One of them is

$B_{50}^{\lor}$ $(((x \lor y)' \lor z)' \lor (x \lor (z' \lor (z \lor u)')')')' = z \,.$

4.4. Boolean Rings and Groups

The ring presentation of Boolean algebras is largely used, due to its computional facilities. After several forerunners (cf. Birkhoff [1948], Ch.X, footnote 3), it is Stone [1935a], [1936] who provided a clear-cut descripion of the relationship between Boolean algebras and Boolean rings, which generalizes the relationship between the two-element Boolean algebra and the ring of integers modulo 2.

Given a Boolean algebra $(B, \vee, \wedge, ', 0, 1)$, the transformation Φ defined by $0 = 0$, $1 = 1$ and

$$(1) \qquad\qquad x + y = (x \wedge y') \vee (x' \wedge y) \,,$$

$$(2) \qquad\qquad xy = x \wedge y \,,$$

produces a *Boolean ring* $(B, +, \cdot, 0, 1)$, that is, a ring with unit in which every element is *idempotent*, i.e., identity $x^2 = x$ holds.

In fact the idempotency implies that *the ring is commutative and of characteristic 2*, meaning that identity $x + x = 0$ holds true. For $(x + y)(x + y) = x + y$, hence $xy + yx = 0$, then $xy = -yx$. Taking $y := x$ we obtain $x = -x$, therefore the previous identity becomes $xy = yx$.

Conversely, a Boolean ring $(B, +, \cdot, 0, 1)$ is made into a Boolean algebra $(B, \vee, \wedge, ', 0, 1)$ by the transformation Ψ defined by $0 = 0$, $1 = 1$ and

$$(3) \qquad\qquad x \vee y = x + y + xy \,,$$

$$(4) \qquad\qquad x \wedge y = xy \,,$$

$$(5) \qquad\qquad x' = x + 1 \,.$$

The above transformations Φ and Ψ satisfy conditions (i)–(iii) in the Introduction of this chapter. Moreover, the above Boolean ring is unique, to the effect that for a given Boolean algebra, any Boolean ring for which there exist polynomial transformations satisfying conditions (i)–(iii) is isomorphic to the one given by (1) and (2); cf. Rudeanu [1961] and Grätzer [1962]. To be specific, the Boolean ring $(B, \oplus, \odot, \alpha, \omega)$ determined on a Boolean algebra $(B, \vee, \wedge, ', 0, 1)$ by Boolean (i.e., polynomial) operations $\oplus, \odot$ are of the form

$$x \oplus y = x + y + a \,,$$

$$x \odot y = xy + a(x + y) \,,$$

and they are isomorphic. See also Rudeanu [1974], Ch.12.

Consider the case when the Boolean algebra is a *field of sets* $(B, \cup, \cap, \mathbf{C}, \varnothing, S)$, meaning that B is a family of subsets of a certain set S, closed with respect to set-theoretical union, intersection and complementation, and $\varnothing, S \in B$. Then the operations of the associated Boolean ring are the symmetric difference of sets and the intersection. If the set S is infinite, then the finite subsets of S form a subring which is still idempotent, hence commutative and of characteristic 2, but has no unit. Such a ring will be called a *generalized Boolean ring*. As shown by Grätzer [1978], there is a Stone-like correspondence between relatively complemented distributive lattices with zero and generalized Boolean rings, $x + y$ being defined as the relative complement of $x \wedge y$ in the segment $[0, x \vee y]$ and, of course, formula (5) is missing. Note that certain authors refer to Boolean rings and generalized Boolean rings as Boolean rings with unit and Boolean rings, respectively.

In the sequel we adopt the convention that $\cdot$ *binds stronger than* $+$. Unless otherwise stated, Boolean rings are characterized within algebras $(B, +, \cdot, 0, 1)$ of type (2,2,0,0) and generalized Boolean rings within algebras $(B, +, \cdot, 0)$ of type (2,2,0).

The axiomatics of Boolean rings began with several systems of axioms for generalized Boolean rings. Thus, Stabler [1941] suggested the system

$$\mathbf{B}_{32}^- = \{ \mathrm{R}_1, \mathrm{R}_2, \mathrm{R}_2', \mathrm{R}_3, \mathrm{R}_4, \mathrm{R}_5, \mathrm{R}_5' \} \ ,$$

where

$\mathrm{R}_1 \qquad x + (y + z) = (x + y) + z$,

$\mathrm{R}_2 \qquad (\exists z \ \ x + z = y + z) \Longrightarrow x = y$,

$\mathrm{R}_2' \qquad (\exists z \ \ z + x = z + y) \Longrightarrow x = y$,

$\mathrm{R}_3 \qquad x^2 = x$,

$\mathrm{R}_4 \qquad x(yz) = (xy)z$,

$\mathrm{R}_5 \qquad x(y + z) = xy + xz$,

$\mathrm{R}_5' \qquad (y + z)x = yx + zx$

and the equational basis

$$\mathbf{B}_{33}^- = \{ \mathrm{R}_3, \mathrm{R}_5, \mathrm{R}_6, \mathrm{R}_7, \mathrm{R}_8 \} \ ,$$

where

$\mathrm{R}_6 \qquad (x + x) + y = y$,

$\mathrm{R}_7 \qquad x + (y + z) = y + (z + x)$,

$\mathrm{R}_8 \qquad x(yz) = y(zx)$,

while Bernstein [1944] provided 9 independent equational bases, for instance

$$\mathbf{B}_{34}^{-} = \{R_3, R_5, R_7, R_8, R_9\} \ ,$$

where

$R_9 \qquad x + (x + y) = y$.

Of course, from every system of axioms $\mathbf{B}^{-}$ for generalized Boolean rings one obtains a basis $\mathbf{B}$ for Boolean rings by adding the axiom $x \cdot 1 = x$ or the axiom $1 \cdot x = x$.

Miller [1952] constructed the independent systems

$$\mathbf{B}_{35} = \{R_{10}, R_{11}, R'_{11}, R_{12}\} \ ,$$

where

$R_{10} \qquad x + (y + y) = x$,

$R_{11} \qquad x \cdot 1 = x$,

$R'_{11} \qquad 1 \cdot x = x$,

$R_{12} \qquad ((x(yy))z)((t + u) + v) = ((zy)x)(v + u) + ((zy)x)t$,

and

$$\mathbf{B}_{36} = \{R_{10}, R_{11}, R_{13}, R_{14}\} \ ,$$

where

$R_{13} \qquad (x(yy))z = (zy)x$,

$R_{14} \qquad x((y + z) + t) = x(t + z) + xy$.

Byrne [1951] obtained the system

$$\mathbf{B}_{37} = \{R_1, R_{15}, R_{16}, R_{17}\} \ ,$$

where

$R_{15} \qquad (x + y) + x = y$,

$R_{16} \qquad x(t + yz) = xt + y(zx)$,

$R_{17} \qquad x(x + 1) = y + y \cdot 1$,

and another system consisting of R_{16} and a rather long identity.

Tamura [1970] devised the system

$$\mathbf{B}_{38} = \{R_{11}, R_{18}, R_{19}, R_{20}\} \ ,$$

where

$R_{18} \qquad x + 0 = x$,

R_{19} $(x + x)y = 0$,

R_{20} $(x + (ys + zt))s = (ys + xs) + t(zs)$,

while Isobe [1973]* discussed the case of generalized Boolean rings (cf. MR 48(1974), #2016).

The additive goup of a Boolean ring led to the concept of a *Boolean group*; by this term is meant a commutative group which satisfies the identity $x + x = 0$. Boolean groups can be characterized by single identities and can be used to obtain short characterizations of Boolean rings.

Thus, Bernstein [1939] provided 20 independent systems of axioms for Boolean groups, among which 12 systems with 2 axioms each, namely

$$\{R_6, R_7\}, \{R_{15}, R_7\}, \{R_{21}, R_7\}, \{R_9, R_7\}, \{R_{10}, R_7\}, \{R_{22}, R_7\},$$

$$\{R_{10}, R_1\}, \{R_6, R_{23}\}, \{R_6, R_{24}\}, \{R_6\,R_{25}\}, \{R_{26}, R_1\}, \{R_{26}, R_7\},$$

where

R_{21} $(x + y) + y = x$,

R_{22} $x + (y + y) = x$,

R_{23} $x + (y + z) = z + (y + x)$,

R_{24} $(x + y) + z = (x + z) + y$,

R_{25} $x + (y + z) = (x + z) + y$,

R_{26} $x + y = x \Longrightarrow y = z + x$.

Let us prove, for instance, that system $\{R_6, R_{24}\}$ defines Boolean groups. We obtain in turn

$$y + z = ((x + x) + y) + z = ((x + x) + z) + y = z + y ,$$

$$(x + y) + z = (y + x) + z = (y + z) + x = x + (y + z) ,$$

$$y + y = (x + x) + (y + y) = (y + y) + (x + x) = x + x ,$$

so that we can set $x + x = 0$ (a constant) and R_6 becomes $0 + y = y$.

Sholander [1953] proved that the identity

R_{27} $x + (y + (z + (y + (z + (y + x))))) = y$

characterizes Boolean groups, while Boolean rings are defined by the system

$$\mathbf{B}_{39} = \{R_{27}, R_{28}\} ,$$

where

R_{28} $x + ((zz)x + (x(y + z) + y(zt))) = yx + ((1 \cdot 1)x + (t(y + y) + z(yt)))$.

Proposition 4.1 and Theorems 4.1 and 4.2 below are due to Mendelsohn and Padmanabhan [1972].

Proposition 4.4.1. *A groupoid* $(B, +)$ *is a Boolean group if and only if it satisfies the identity*

R_{29} $x + (((x + y) + z) + y) = z$.

PROOF: Setting

$$y := ((x + t) + u) + t, \; z := ((u + v) + x) + v$$

and using R_{29}, we have $x + y = u$ and $u + z = x$, hence $((x + y) + z) + y = x + y = u$, so that R_{29} becomes

(6) $x + u = ((u + v) + x) + v$,

whence by pre-adding u and using again R_{29} we get

(7) $u + (x + u) = x$.

Now it follows from R_{29} and (7) that

(8) $z = x + (((x + x) + z) + x) = (x + x) + z$,

hence $z + z = z + ((x + x) + z) = x + x$, again by (7). Therefore we can set $x + x = 0$ (a constant) and property (8) reads $z = 0 + z$, hence (7) implies $x = 0 + (x + 0) = x + 0$.

Thus 0 is element zero with respect to addition, so that identity $x + x = 0$ shows that $-x$ exists for every x, namely $-x = x$. Finally (6) and (7) imply in turn

$$x + u = ((u + 0) + x) + 0 = u + x ,$$

$$(x + u) + v = (((u + v) + x) + v) + v = v + (((u + v) + x) + v)$$

$$= (u + v) + x = x + (u + v) .$$

□

Mendelsohn and Padmanabhan [1975] proved also that Boolean groups have exactly six single-identity definitions of shortest length, one of them being R_{29}.

Theorem 4.4.1. *Let B be an algebra whose signature Σ contains a distinguished binary operation, say $+$. Let $x, y, z, x_1, x_2, \ldots, x_n$ be variables subject to the condition $x, y, z \notin \{x_1, \ldots, x_n\}$, and $w(x_1, \ldots, x_n)$ a word of signature Σ. Then the reduct $(B, +)$ is a Boolean group and the algebra B satisfies the identity $w(x_1, \ldots, x_n) = 0$ if and only if it satisfies the identity*

$$(9) \qquad x + ((((x + y) + w) + z) + y) = z \ .$$

PROOF: Setting $y := (((x+t)+w)+u)+t$ in (9) and using (9), we obtain $x + y = u$, so that (9) becomes

$$(10) \qquad x + (((u + w) + z) + ((((x + t) + w) + u) + t)) = z \ .$$

Setting $z := ((((u + w) + v) + w) + x) + v$ in (10) and using (9), we have $(u + w) + z = x$, so that the left-hand side of (10) becomes $x + (x + ((((x + t) + w) + u) + t)) = x + u$ by (9), therefore (10) reduces to

$$(11) \qquad x + u = ((((u + w) + v) + w) + x) + v \ .$$

Setting $x := (u + w) + z$ in (10) and using (9), the left-hand side of (10) becomes $x + (x + (((x + t) + w) + u) + t) = x + u$ by (9), therefore (10) reduces to

$$(12) \qquad ((u + w) + z) + u = z \ .$$

Setting $u := w$ in (11) and taking into account (12) we get

$$(13) \qquad x + w = ((((w + w) + v) + w) + x) + v = (v + x) + v \ .$$

Setting $x := w$ in (9) and using (13) and (12) we obtain

$$(14) \qquad \begin{aligned} z &= w + ((((w + y) + w) + z) + y) \\ &= w + (((y + w) + z) + y) = w + z \ . \end{aligned}$$

Setting $u := w$ in (11) and using (14) and (12) yields

$$x + w = ((((w + w) + v) + w) + x) + v$$

$$= (((w + v) + w) + x) + v = ((v + w) + x) + v = x$$

and since $x + w = x$ we see that (9) reduces to R_{29}, therefore $(B, +)$ is a Boolean group by Proposition 4.1. Now taking $x := y := z := 0$ in (9) we obtain the identity $w = 0$. $\qquad\square$

Theorem 4.4.2. *An algebra* $(B, +, \cdot, 1)$ *of type* $(2,2,0)$ *is a Boolean ring if and only if it satisfies the single identity* (9) *with*

$$w = x_1^2 + x_1 + x_2(x_3 + x_4) + x_2 x_3 + x_2 x_4 + x_5 \cdot 1 + x_5 + (x_6 x_7)x_8 + (x_7 x_8)x_6 \ .$$

PROOF: Necessity is immediate. To prove sufficiency, note first that $(B, +)$ is a Boolean group by Theorem 4.1. In particular the operation $+$ is associative, so that w makes sense.

Identity $w = 0$ with all $x_i := 0$ except x_5 yields $x_5 \cdot 1 + x_5 = 0$, that is, $x_5 \cdot 1 = x_5$. Hence, setting $x_1 := 1, x_6 := x_7 := x_8 := 0$ in $w = 0$ we obtain $x_2(x_3 + x_4) = x_2 x_3 + x_2 x_4$. Thus $w = 0$ reduces to

$$x_1^2 + x_1 + (x_6 x_7)x_8 + (x_7 x_8)x_6 = 0$$

and puting again $x_6 := x_7 := x_8 := 0$ we obtain the idempotency $x_1^2 = x_1$. Therefore $(x_6 x_7)x_8 = (x_7 x_8)x_6$, and finally this implies $x_6 x_8 = x_8 x_6$ by taking $x_7 := 1$. $\qquad\qquad\square$

Other equational bases use the ring operations plus an extra unary operation. Wooyenaka [1964] characterized Boolean rings in terms of the operations $+, \cdot$ and $'$. Morgado [1970] used the extra operation $-$ (unary). By simplifying a system due to Iséki [1968], Morgado characterized Boolean rings by the system

$$\mathbf{B}_{40} = \{R_1', R_{18}, R_{30}, R_{31}\} \ ,$$

where

R_{30} $\qquad (-x + x)y = 0 \ ,$

R_{31} $\qquad ((xt + yt) + zs)s = y(st) + (x(ts) + zs) \ ,$

and generalized Boolean rings by the system

$$\mathbf{B}_{41}^- = \{R_3, R_{18}', R_{30}, R_{32}\} \ ,$$

where

R_{18}' $\qquad 0 + x = x \ ,$

R_{32} $\qquad ((xt + y) + z)s = sy + (x(ts) + zs) \ .$

Stabler [1941] proved that the generalized Boolean rings can also be characterized in terms of the operations $\vee, +$ by the system

$$\mathbf{B}_{42}^- = \{R_2, R_2', R_{33}, R_{34}\} \ ,$$

where

R_{33} $\qquad x + ((x \vee y) + (x \vee z)) = x \vee (y + z) \ ,$

R_{34} $\qquad ((x \vee z) + (y \vee z)) + z = (x + y) \vee z \ .$

4.5. Boolean Algebras in Terms of Nonassociative Binary Operations

The nonassociative binary operations dealt with in this section are the *difference* $x - y = x \wedge y'$, also known as *exception* operation (*"x* but not *y"*), the *implication* $x \to y = x' \vee y$ (*"if x* then *y"*), the *Sheffer stroke* $x \mid y = x' \vee y'$ and the *rejection* or *Peirce operation* $x \mid\mid y = x' \wedge y'$ (*"neither x* nor *y"*); the last two operations are widely used in computer science under the names NAND (*"not and"*) and NOR (*"not or"*), respectively.

Boolean algebras in terms of difference are presented as algebras $(B, -, 1)$ of type $(2,0)$, the functions Ψ and Φ referred to in the beginning of this chapter being

(1) $$x' = 1 - x \,,$$

(2) $$x \wedge y = x - y' = x - (1 - y) \,,$$

(3) $$x \vee y = (x' - y)' \,,$$

(4) $$0 = 1' = x - x \,,$$

and $\Phi : x - y = x \wedge y'$, $1 = 1$. It is readily checked that $\Psi\Phi(F) = F$, while for $\Phi\Psi(G) = G$ the system of axioms should imply $x - (1 - (1 - y)) = x - y$.

The presentations in terms of implication exhibit algebras $(B, \to, 0)$ of type $(2,0)$, with Ψ given by

(5) $$x' = x \to 0 \,,$$

(6) $$x \vee y = x' \to y = (x \to y) \to y \,,$$

(7) $$x \wedge y = (x \to y')' \,,$$

(8) $$1 = 0' = x \to x \,,$$

and $\Phi : x \to y = x' \vee y$, $0 = 0$. Same comment as above about condition (iii).

It is well known that Boolean algebras satisfy the principle of duality, where Boolean duality acts as lattice duality $\vee \leftrightarrow \wedge$, $\leq \leftrightarrow \geq$, $0 \leftrightarrow 1$, and leaves the complementation $'$ invariant. Note that $x - y$ and $y \to x$ are dual to each other; we can say that $-$ and $\to$ are *skew dual* to each other. Moreover, this remark extends to formulae (2)–(4) with respect to formulae (6)–(8); for instance, the skew dual of (2) is $y \vee x = y' \to x$, which coincides with (6); etc. A practical consequence of this remark is that *every system of axioms in terms of difference can be translated into a system of axioms in terms of implication, and conversely.* In the following we use exclusively the language of difference and whenever the original system was in terms of implication we mention this fact.

Unless otherwise stated, it is understood that any term of the form a' ocurring in an axiom is a shorthand for $1 - a$.

The first system of axioms in terms of difference was given by Bernstein [1914]*, namely the independent system

$$\mathbf{B}_{43} = \{B_{51}, B_{52}, B_{53}, B_{54}\} \, ,$$

where

B_{51} $x - (y - x) = x \, ,$

B_{52} $(x - x) - z = (y - y) - t \, ,$

B_{53} $x - y = y - x \Longrightarrow x = y \, ,$

B_{54} $(x - y) - (z - t) = (((x - y) - t')' - ((z' - x') - y))' \, .$

The simpler system

$$\mathbf{B}_{44} = \{B_{55}, B_{56}, B_{57}\} \, ,$$

where

B_{55} $x - (y - y) = x \, ,$

B_{56} $x - y' = y - x' \, ,$

B_{57} $((x - y)' - (x - z'))' = x - (y - z) \, ,$

was devised by Taylor [1920a], together with its complete existential theory.

The original form of the following four systems is in terms of implication:

$$\mathbf{B}_{45} = \{B_{58}, B_{59}, B_{60}, B_{61}, B_{62}\}$$

provided by Huntington [1932], where

B_{58} $x - 1 = x - x \, ,$

B_{59} $x - (x - y) = y - (y - x) \, ,$

B_{60} $(x - (x - y)) - ((x - (x - y)) - z) = x - (x - (y - (y - z))) \, ,$

B_{61} $(x - y)' - ((x - y)' - (x' - y)')) = y \, ,$

B_{62} $x - (x - y') = x - y \, ,$

the independent system

$$\mathbf{B}_{46} = \{B_{51}, B_{63}\} \, ,$$

due to Bernstein [1934], where

B_{63} $((x - y') - (z - x'))' = (x - (y - z)) - (t - t)$,

and the independent systems

$$\mathbf{B}_{47} = \{B_{55}, B_{64}\} \ ,$$

$$\mathbf{B}_{48} = \{B_{65}, B_{66}\} \ ,$$

devised by Diego and Suarez [1966], where

B_{64} $(x' - y') - z = (y - z) - (x - z)$,

B_{65} $x'' - (y - y) = x$,

B_{66} $(x' - y')'' - z = (y'' - z) - (x'' - z)$.

Güting [1971] simplified system B_{32}, but he was unaware of systems $\mathbf{B}_{33} - \mathbf{B}_{37}$. He suggested the system

$$\mathbf{B}_{49} = \{B_{67}, B_{68}, B_{69}, B_{70}\} \ ,$$

where

B_{67} $(x - y) - z = (x - z) - y$,

B_{68} $1 - (1 - x) = x$ (B_9 would be a shorthand) ,

B_{69} $x - x = 1'$,

B_{70} $x - (x - y) = x - y'$.

The next result provides a system in which $-$ and $'$ are basic operations, while 0 and 1 are constructed:

Proposition 4.5.1. (Taylor [1920a]) *The following system characterizes Boolean algebras within algebras* $(B, -, ')$ *of type* (2,1):

$$\mathbf{B}_{50} = \{B_{55}, B_{71}, B_{72}\} \ ,$$

where

B_{71} $x - x' = x$,

B_{72} $((z - x')' - (y' - x'))' = x - (y - z)$.

PROOF: B_{72} and B_{55} imply $((x - x')' - (x' - x'))' = x - (x - x) = x$, while from B_{55} and B_{71} we get $((x - x')' - (x' - x'))' = (x - x')'' = x''$, therefore $x'' = x$. Now B_{71} yields $x' - x = x' - x'' = x'$.

Using in turn $a'' = a$, $b' - b = b'$, B_{72} and B_{71}, we obtain

$$y' - x' = (y' - x')'' = ((y' - x')' - (y' - x'))' = x - (y - y') = x - y \ ,$$

which, due to $a'' = a$, implies also $x' - y = y' - x$ and $x - y' = y - x'$. Now

$$(x - x)' = (x - x)' - (y - y) = (y - y)' - (x - x) = (y - y)' ,$$

therefore we can set $(x - x)' = (y - y)' = 1$. Then

$$1 - x = (x - x)' - x = x' - (x - x) = x' ,$$

hence the algebra $(B, -, 1)$ satisfies axioms B_{55} and B_{56} (recall that in B_{56}, x' is used as a shorthand for $1 - x$). If we succeed to prove B_{57}, it will follow that the algebra $(B, -, 1)$ satisfies system $\mathbf{B}_{44}$, hence it is Boolean. But

$$(z - x')' - (y' - x') = (x - z')' - (x - y) = (x - y)' - (x - z') ,$$

so that axiom B_{72} can be written in the form B_{57}.

Finally note that $\Phi : x - y = x \wedge y'$, $x' = x'$, $\Psi : (2), (3), (4), x' = x'$, and conditions (i),(iii) are readily checked.   □

Stone [1935b] and Sampathkumar [1967]* (cf. MR 50(1975),#12846) defined Boolean algebras by using the restriction of the difference $x - y$ to the case when $y \leq x$.

Iséki [1965a], [1965b], [1965c], [1972], Arai and Iséki [1965], Sicoe [1966], Imai and Iséki [1976], characterized Boolean algebras in terms of the difference $x - y$ and the partial order $\leq$.

Bosbach [1969] provided an axiom system in terms of the difference $x - y$ and the join $x \vee y$. See also Bosbach [1977].

Boolean algebras in terms of the Sheffer stroke $|$ are algebras $(B, |)$ of type (2), the transformation Ψ being

(9) $$x' = x \mid x ,$$

(10) $$x \vee y = (x \mid x) \mid (y \mid y) ,$$

(11) $$x \wedge y = (x \mid y) \mid (x \mid y) ,$$

(12) $$1 = x \mid (x \mid x) ,$$

(13) $$0 = (x \mid (x \mid x)) \mid (x \mid (x \mid x)) ,$$

and $\Phi : x \mid y = x' \vee y'$, while for the algebras $(B, \|)$ of type (2) the transformation Ψ is

(14) $$x' = x \parallel x ,$$

(15) $$x \wedge y = (x \,\|\, x) \,\|\, (y \,\|\, y) \,,$$

(16) $$x \vee y = (x \,\|\, y) \,\|\, (x \,\|\, y) \,,$$

(17) $$0 = x \,\|\, (x \,\|\, x) \,,$$

(18) $$1 = (x \,\|\, (x \,\|\, x)) \,\|\, (x \,\|\, (x \,\|\, x)) \,,$$

and $\Phi : x \,\|\, y = x' \wedge y'$. So the usual duality holds between (B, M) with (9)–(13) and $(B, \|)$ with (14)–(18).

There is a pre-history of this approach, due to Stamm [1911], who constructed the system

$$\mathbf{B}_{51} = \{\mathrm{B}_{73}^{|}, \mathrm{B}_{73}^{\|}, \mathrm{B}_{74}^{|}, \mathrm{B}_{74}^{\|}, \mathrm{B}_{75}, \mathrm{B}_{76}^{|}, \mathrm{B}_{77}^{|}, \mathrm{B}_{78}^{|}\} \,,$$

where

$\mathrm{B}_{73}^{|}$ $x \mid y = y \mid x \,,$

$\mathrm{B}_{73}^{\|}$ $x \,\|\, y = y \,\|\, x \,,$

$\mathrm{B}_{74}^{|}$ $(x \mid x) \mid (y \mid z) = (x \,\|\, y) \,\|\, (x \,\|\, z) \,,$

$\mathrm{B}_{74}^{\|}$ $(x \,\|\, x) \,\|\, (y \,\|\, z) = (x \mid y) \mid (x \mid z) \,,$

B_{75} $x \mid x = x \,\|\, x \,,$

$\mathrm{B}_{76}^{|}$ $x \mid (x \mid x) = y \mid (y \mid y) \,,$

$\mathrm{B}_{77}^{|}$ $x \mid (y \mid (y \mid y)) = x \mid x \,,$

$\mathrm{B}_{78}^{|}$ $(x \mid x) \mid (x \mid x) = x \,.$

Sheffer [1913] provided the independent system

$$\mathbf{B}_{52} = \{\mathrm{B}_{77}^{|}, \mathrm{B}_{78}^{|}, \mathrm{B}_{79}^{|}\} \,,$$

where

$\mathrm{B}_{79}^{|}$ $(x \mid (y \mid z)) \mid (x \mid (y \mid z)) = ((y \mid y) \mid x) \mid ((z \mid z) \mid x) \mid x) \,,$

while Dines [1914/5] proved that this system is not completely independent.

Bernstein [1916] devised the independent 2-basis

$$\mathbf{B}_{53} = \{\mathrm{B}_{80}^{|}, \mathrm{B}_{81}^{|}\} \,,$$

where

$\mathrm{B}_{80}^{|}$ $(y \mid x) \mid ((y \mid y) \mid x) = x \,,$

$\mathrm{B}_{81}^{|}$ $((y \mid (x \mid x)) \mid ((z \mid z) \mid (x \mid x))) \mid ((y \mid (x \mid x)) \mid ((z \mid z) \mid (x \mid x)))$
     $= (x \mid x) \mid ((y \mid y) \, z)) \,.$

Curiously enough, if one assumes that the support B has at least 4 elements, then the systems $\mathbf{B}_{52}$ and $\mathbf{B}_{53}$ become completely independent, as was proved by Taylor [1920b] and [1917], respectively (in the latter paper it is also proved that $\mathbf{B}_{53}$ is not completely independent).

Other two-identity systems were given by Bernstein [1933], who constructed the basis

$$\mathbf{B}_{54} = \{\mathbf{B}^{|}_{80}, \mathbf{B}^{|}_{82}\} \,,$$

where

$\mathbf{B}^{|}_{82}$ $(((z \mid z) \mid x) \mid ((y \mid y) \mid x)) \mid (((z \mid z) \mid x) \mid ((y \mid y) \mid x)) = x \mid (y \mid z) \,,$

Meredith [1969], who obtained the system

$$\mathbf{B}_{55} = \{\mathbf{B}^{|}_{83}, \mathbf{B}^{|}_{84}\} \,,$$

where

$\mathbf{B}^{|}_{83}$ $(x \mid x) \mid (y \mid x) = x \,,$

$\mathbf{B}^{|}_{84}$ $x \mid (y \mid (x \mid z)) = ((z \mid y) \mid y) \mid x \,,$

and Veroff [2000], who devised the system

$$\mathbf{B}_{56} = \{\mathbf{B}^{|}_{73}, \mathbf{B}^{|}_{85}\} \,,$$

where

$\mathbf{B}^{|}_{85}$ $(x \mid y) \mid (x \mid (y \mid z)) = x \,.$

There are also single-axiom characterizations of Boolean algebras in terms of Sheffer stroke.

Thus, Hoberman and McKinsey [1937] proved that *an algebra (B, M) can be made into a Boolean algebra if and only if every polynomial function $f : B \longrightarrow B$ can be written in the form $f(x) = (f(1) \wedge x) \vee (f(1') \wedge x')$, where $', \vee, \wedge, 1$ are given by formulae* (9)–(12).

Sholander [1953] characterized Boolean algebras by the single axiom

$\mathbf{B}^{|}_{86}$ $(x \mid (y' \mid y))'' = (x \mid (y' \mid z))'' \implies (y \mid x) \mid (z' \mid x) = x \,,$

where a' is a shorthand for $a \mid a$.

McCune, Veroff, Fittelson, Harris,, Feist and Wos [2002] provided an Otter-assisted proof of the fact that the single identity

$\mathbf{B}^{|}_{87}$ $(x \mid ((y \mid x) \mid x)) \mid (y \mid (z \mid x)) = y$

defines Boolean algebras. They also proved that there is no shorter single identity defining Boolean algebras in terms of the Sheffer stroke.

Conjecture $B_{87}^|$ has the shortest possible length (15 symbols) among the single identities defining Boolean algebras under any treatment.

We conclude by emphasizing that, beside nonassociativity, there is one more link between the three operations dealt with in this section.

Note first that, in view of duality, any system of axioms in terms of the difference $x - y$/implication $x \to y$ yields a system of axioms in terms of $x \vee y' = y \to x$/of $x' \wedge y = y - x$. Therefore the functions

$$(19) \qquad x \mid y, \ x \parallel y, \ x - y, \ y - x, \ x \to y, \ y \to x$$

share the property that each of them is a binary simple Boolean function (i.e., a Boolean polynomial involving no constants) which can be used alone to define Boolean algebras as described in the Introduction of this chapter. For it is clear that in the case of the functions $\mid$ and $\parallel$ the signature Θ consists of the function alone, but this is also true for the remaining four functions (19) provided we incorporate the constant 0 or 1 into the corresponding transformation Φ; for instance, in the case of the function $x - y$, x' is the Θ−polynomial $1 - x$; etc.

The functions (19) are the only functions with the above property. The reason is that, as was proved by Rudeanu [1961], the functions (19) are the only binary simple Boolean functions f such that every Boolean function (i.e., Boolean polynomial) is an f−polynomial. Besides, these f−polynomials involve no constants only in the case of the functions $\mid$ and $\parallel$. The latter result was first proved by Zylinski [1925] and Lalan [1950].

Thus all the functions (19) might be called generalized Sheffer functions. The paper by Rudeanu [1961] studies also the even more general case when the Sheffer functions are not required to be simple Boolean functions, but just Boolean functions. See also Rudeanu [1974], Ch.12.

4.6. Boolean Algebras in Terms of Ternary or *n*-ary Operations

In this section we survey axiomatizations of Boolean algebras in terms of ternary majority decision, ternary rejection, conditional disjunction and of a generalization of the Sheffer stroke to n variables.

We have seen in Chapter 3, §3, that bounded distributive lattices can be characterized in terms of the median operation

$$(1) \qquad m(x, y, z) = (y \wedge z) \vee (x \wedge z) \vee (x \wedge y) = (y \vee z) \wedge (x \vee z) \wedge (x \vee y) .$$

This self-dual operation, also known as *ternary majority decision*, yields the lattice operations via formulae

$$(2) \qquad\qquad x \vee y = m(x,1,y) \,, \quad x \wedge y = m(x,0,y) \,.$$

In particular in a Boolean algebra we have

$$(3) \qquad (y' \wedge z') \vee (x' \wedge z') \vee (x' \wedge y') = (y' \vee z') \wedge (x' \vee z') \wedge (x' \vee y') \,,$$

which reads

$$(4) \qquad\qquad m(x',y',z') = m'(x,y,z) \,,$$

and the function m' is known as *ternary rejection*.

It follows by property $x'' = x$ that any system of axioms in terms of majority decision can be translated into an axiom system using ternary rejection, and conversely. This explains the resemblance between the existing systems of the two kinds.

Clearly, in order to define Boolean algebras in terms of the majority decision m we also need the operation $'$. Although formulae (2) involve the constants 0 and 1, one of them can be dispensed with in the signature, for instance one defines $1 = 0'$. Moreover, Grau [1947] provided a system of axioms for Boolean algebras involving only the operations m and $'$ without any constant, namely

$$\mathbf{B}_{57} = \{\mathrm{B}_{88}, \mathrm{B}_{89}, \mathrm{B}_{90}, \mathrm{B}_{91}, \mathrm{B}_{92}\} \,,$$

where

$\mathrm{B}_{88} \qquad m(x,y,m(z,t,u)) = m(m(x,y,z),t,m(x,y,u)) \,,$

$\mathrm{B}_{89} \qquad m(x,y,y) = y \,,$

$\mathrm{B}_{90} \qquad m(y,y,x) = y \,,$

$\mathrm{B}_{91} \qquad m(x,y,y') = x \,,$

$\mathrm{B}_{92} \qquad m(y',y,x) = x \,.$

Instead of (2), Grau chooses an element p, defines

$$(5) \qquad\qquad x \vee y = m(x,p',y) \,, \quad x \wedge y = m(x,p,y) \,,$$

and proves that the resulting algebra is a Boolean algebra $B(p)$ having p and p' as zero and one, respectively. Moreover, the algebras $B(p)$ and $B(q)$ are isomorphic for any p and q. (As a matter of fact, this isomorphism is not limited to the algebras constructed by Grau, but it holds for all the

Boolean algebras defined by polynomials on a given Boolean algebra, as was proved by Goetz [1971]; see also Rudeanu [1974], Ch.12, §4).

Kalicki [1952]* proved that axiom B_{90} follows from the other axioms of $\mathbf{B}_{57}$, while the system

$$\mathbf{B}_{58} = \{B_{88}, B_{89}, B_{91}, B_{92}\}$$

is independent.

We have seen in the previous chapters that changing the "natural" order of the variables in a certain axiom may result in the creation of a shorter axiom system. Likewise, changing the order of the variables in B_{88} produces a two-axiom system for Boolean algebras in terms of m and $'$:

Proposition 4.6.1. (Croisot [1951]) *The following system chracterizes Boolean algebras within algebras* $(B, m, ')$ *of type* $(3,1)$:

$$\mathbf{B}_{59} = \{B_{93}, B_{94}\} \, ,$$

where

B_{93} $m(x, y, m(z, t, u)) = m(z, m(y, x, u), m(x, y, t))$,

B_{94} $m(y, x, y') = x$.

PROOF: We have

$$\begin{aligned}
(5') \qquad m(x, y, x) &= m(x, m(x, y, x'), m(y, x, y')) \\
&= m(y, x, m(x, y', x')) = m(y, x, y') = x \, ,
\end{aligned}$$

hence

$$\begin{aligned}
(6) \qquad m(y, x, x) &= m(y, x, m(x, y, x)) \\
&= m(x, m(x, y, x), m(y, x, y)) = m(x, x, y) \, ,
\end{aligned}$$

so that we can prove B_{89} by $(5')$, B_{93}, (6), B_{93}, $(5')$ and $(5')$:

$$m(y, x, x) = m(y, m(x, x, x), m(x, x, x)) = m(x, x, m(y, x, x))$$

$$= m(x, x, m(x, x, y)) = m(x, m(x, x, y), m(x, x, x)) = m(x, m(x, x, y), x) = x \, .$$

Now B_{89}, B_{93}, (6) and B_{89} imply

$$\begin{aligned}
(7) \qquad m(x, y, z) &= m(x, t, m(t, z, z)) \\
&= m(t, m(y, x, z), m(x, y, z)) = m(m(x, y, z), t, t) = m(y, x, z)
\end{aligned}$$

provided we take $t := m(y, x, z)$. Further we use B$_{94}$, (5$'$), (7), B$_{93}$, (7), B$_{94}$, B$_{94}$ and obtain

$$m(y', x, y) = m(y', m(y, x, y'), m(y, x, y)) = m(y', m(x, y, y'), m(y, x, y))$$

$$= m(y, x, m(y', y, y')) = m(y, x, m(y, y', y')) = m(y, x, y') = x \ ,$$

that is,

$$(8) \qquad\qquad\qquad m(y', x, y) = x \ .$$

Then we obtain B$_{92}$, by B$_{89}$, B$_{94}$, B$_{93}$, (8) and B$_{94}$:

$$m(y', y, x) = m(y', m(x, y, y), m(y, x, y'))$$

$$= m(y, x, m(y', y', y)) = m(y, x, y') = x \ .$$

Finally we apply B$_{92}$, B$_{93}$, (7), B$_{92}$, B$_{92}$ and obtain

$$(9) \qquad \begin{aligned} m(x, y, z) &= m(t', t, m(x, y, z)) = m(x, m(t, t', z), m(t', t, y)) \\ &= m(x, m(t', t, z), y) = m(x, z, y) \ . \end{aligned}$$

It follows from (7) and (9) via Lemma 3.3.1 that m is invariant to any permutation of the variables. Therefore identity (8) reduces to B$_{91}$ and B$_{93}$ yields B$_{88}$ because

$$m(x, y, m(z, t, u)) = m(x, y, m(t, z, u)) = m(t, m(y, x, u), m(x, y, z))$$

$$= m(m(x, y, z), t, m(x, y, u)) \ ,$$

so that system **B**$_{58}$ is fulfilled. $\square$

Note that if we wish to obtain a definitional equivalence as described in the beginning of this section, then we must ensure property $\Phi\Psi = \mathbf{1}$, which amounts to the following identity:

$$(10) \qquad m(m(x, m(y, z, p'), p), m(y, z, p), p') = m(x, y, z) \ .$$

Padmanabhan and McCune [1995] obtained 43 single-identity characterizations of Boolean algebras in terms of the operations m and $'$. One of them was constructed using a technique due to Padmanabhan and Quackenbush [1973], which will be described in Theorem 9.3 of this book. The other identities were produced with the aid of the computer program Otter, the shortest ones having 7 variables and 26 symbols each. Here they are:

B$_{95}$ $\quad m(m(x, x', y), m(m(z, u, v), w, m(z, u, t)))', m(u, m(u, m(t, w, v), z))$
$\qquad = y \ ,$

B_{96} $m(m(x, x'y), (m(z, m(u, v, w), t))', m(m(t, z, v), v, m(t, z, u))) = y$,

B_{97} $m(m(x, x'y), (m(z, m(u, v, w), t))', m(m(t, z, u), v, m(t, z, w))) = y$.

The definition of Boolean algebras in terms of ternary rejection m' does not require any more the complementation $'$ as a basic operation. To see this, let us re-denote rejection, say $m' = \mu$. Then in every Boolean algebra we have

(11) $\mu(x, y, z) = (y' \wedge z') \vee (x' \wedge z') \vee (x' \wedge y')$,

(12) $x' = \mu(x, x, x)$,

and using (12), formulae (2) can be written in the form

(13) $x \vee y = \mu'(x, 1, y)$, $x \wedge y = \mu'(x, 0, y)$.

Frink [1926] established a definitional equivalence between Boolean algebras $(B, \vee, \wedge, ', 0, 1)$ and algebras $(B, \mu, 0, 1)$ via the transformations Φ defined by (11) and Ψ defined by (12), (13):

Proposition 4.6.2. *The following system characterizes Boolean algebras within algebras $(B, \mu, 0, 1)$ of type (3,0,0):*

$$\mathbf{B}_{60} = \{\text{B}_9, \text{B}_{98}, \text{B}_{99}, \text{B}_{100}, \text{B}_{101}, \text{B}_{102}, \text{B}_{103}\} \ ,$$

whose axioms, using the shortcut (12), *are*

B_{98} $\mu(x, y, z) = \mu(y, x, z)$,

B_{99} $\mu(x, y, z) = \mu(y, z, x)$,

B_{100} $\mu(x, x, y) = \mu(x, x, z)$,

B_{101} $\mu'(x, x', y) = y$,

B_{102} $\mu'(x, \mu'(y, z, t'), t) = \mu'(\mu'(x, y, t), \mu'(x, z, t), t')$,

B_{103} $\mu(x, y, z) = \mu'(\mu'(x', \mu'(y', z', 1), 0), \mu'(y', z', 0), 1)$.

PROOF: Properties (11)–(13) and B_{98}–B_{103} of a Boolean algebra are easy to check. Therefore $\Phi\Psi = 1$.

Now suppose system $\mathbf{B}_{60}$ is fulfilled. Since $\mu(x, y, z) = \mu(y, x, z) = \mu(x, z, y)$ by B_{98} and B_{99}, the hypotheses of Lemma 3.1 are satisfied, hence μ i invariant under any permutation of the variables, therefore so is μ'. Consequently, the join and meet defined by (13) satisfy the commutativity B_2. Besides, the identity $\mu'(0, 1, y) = y$, which is an instance of B_{101}, implies $\mu'(1, 0, y) = y$, showing that identities B_6 hold: $0 \vee y = y$ and $1 \wedge y = y$. Likewise, B_{101} implies $\mu'(x, x', 0 = 0$ and $\mu'(x, x', 1) = 1$, which can be read in the form B_8: $x \wedge x' = 0$ and $x \vee x' = 1$. Furthermore, by applying B_{102} with $t := 0$ we obtain

$$x \wedge (y \vee z) = \mu'(x, \mu'(y, z, 1), 0) = \mu'(\mu'(x, y, 0), \mu'(x, z, 0), 1) = (x \wedge y) \vee (x \wedge z) \ ,$$

that is $B_5^\wedge$, while for $t := 1$ we get $B_5^\vee$. We have thus verified all the axioms of system $\mathbf{B}_2$.

Therefore the transformation Ψ is well defined. Property $\Phi \Psi = \mathbf{1}$ follows by B_{103}, which is a translation of (10). $\square$

The Grau-like approach applies also to ternary rejection, the role of formulae (5) being played by

$$(14) \qquad\qquad x \vee y = \mu'(x, p', y) \ , \quad x \wedge y = \mu'(x, p, y) \ ,$$

where $'$ is the shorthand (12).

Proposition 4.6.3. (Whiteman [1937]) *The following independent system characterizes Boolean algebras within algebras (B, μ) of type (3):*

$$\mathbf{B}_{61} = \{B_{99}, B_{104}, B_{105}\} \ ,$$

where, using the shorthand (12), we have set

$B_{104} \qquad \mu(x', y, y') = x \ ,$

$B_{105} \qquad \mu(x, y, \mu'(z, t, u)) = \mu(\mu'(x, y, z), \mu'(x, y, t), u) \ .$

PROOF: We are going to check all the axioms of Frink's system $\mathbf{B}_{60}$ but B_{103}. This will establish the desired equivalence (but not a definitional equivalence).

We already have B_{99}. Using B_{104} and B_{99} we obtain B_9:

$$x'' = \mu(x''', x', x'') = \mu(x', x'', x''') = x \ .$$

From B_9 and B_{104} we deduce

$$(15) \qquad\qquad \mu(x, y, y') = \mu(x'', y, y') = x' \ .$$

Now B_9, (15) and B_{99} imply

$$x = x'' = \mu'(x, y, y') = \mu(y, y', x) \ ,$$

that is B_{101}. Further note that B_{99}, (15) and B_9 imply

$$(16) \qquad\qquad \mu(x', x', x) = \mu(x', x, x') = x'' = x \ ,$$

while from (15) in the form $\mu'(x, y, y') = x$, B_{105}, (16) and (15) we infer

$$(17) \qquad \mu(x, x, y) = \mu(\mu'(x, y, y'), \mu'(x, y, y'), y)$$
$$= \mu(x, y, \mu'(y', y', y)) = \mu(x, y, y') = x' \, ,$$

which implies B_{100}. Now a successive application of B_{99}, B_{105} and B_{99} yields

$$\mu(x, \mu'(y, z, t'), t) = \mu(t, x, \mu'(y, z, t')) = \mu(\mu'(t, x, y), \mu'(t, x, z), t')$$

$$= \mu(\mu'((x, y, t), \mu'(x, z, t), t') \, ,$$

which implies B_{102}.

Finally we are going to prove B_{98} in several steps. First we observe that B_{99}, (17) and B_9 imply

$$\mu'(z, t, z) = \mu'(z, z, t) = z'' = z \, ;$$

using this, together with B_{105} and B_{99}, we obtain

$$\mu(x, y, z) = \mu(x, y, \mu'(z, t, z)) = \mu(\mu'(x, y, z), \mu'(x, y, t), z)$$

$$= \mu(z, \mu'(x, y, z), \mu'(x, y, t)) = \mu(\mu'(z, \mu'(x, y, z), x), \mu'(z, \mu'(x, y, z), y), t) \, ,$$

showing that $\mu(x, y, z)$ is of the form

$$(18) \qquad \mu(x, y, z) = \mu(\mu'(z, \mu'(x, y, z), x), w, t) \, .$$

Then, using B_{99}, (17) and B_9 in the form

$$(19) \qquad \mu'(a, b, a) = \mu'(b, a, a) = \mu'(a, a, b) = a \, ,$$

we compute as above

$$\mu(z, \mu'(x, y, z), x) = \mu(x, z, \mu'(x, y, z)) = \mu(\mu'(x, z, x), \mu'(x, z, y), z)$$

$$= \mu(x, \mu'(x, z, y), z) = \mu(z, x, \mu'(x, z, y))$$

$$= \mu(\mu'(z, x, x), \mu'(z, x, z), y) = \mu(x, z, y) \, .$$

We introduce this result into (18) with $t := \mu'(x, z, y)$ and use again (19):

$$\mu(x, y, z) = \mu(\mu'(x, z, y), w, t) = \mu(t, w, t) = t' = \mu(x, z, y) = \mu(y, x, z) \, .$$

We omit the proof of independence. $\qquad\qquad \square$

Another ternary Boolean operation is the *conditional disjunction* introduced by Church, namely

(20) $x(y, z) = (x \wedge y) \vee (x' \wedge z)$,

to be read "if x then y else z", which is important in switching theory and computer programming. The fact that Boolean algebras can be defined in terms of conditional disjunction follows from the identities

(21) $x \vee y = x(1, y)$, $x \wedge y = x(y, 0)$, $x' = x(0, 1)$.

It was Dicker [1963] who devised the independent system

$$\mathbf{B}_{62} = \{B_{106}, B_{107}, B_{108}, B_{109}, B_{110}\} ,$$

where

B_{106} $x(y(u, v), z(u, v)) = (x(y, z))(u, v)$,

B_{107} $\exists 0 \ 0(y, x) = x$,

B_{108} $\exists 0 \ x(y, 0) = y(x, 0)$,

B_{109} $\exists 0 \ x(0, x) = 0$,

B_{110} $1(y, x) = y$.

We only note here that the element 0 in axioms B_{106}–B_{110} is unique. For suppose $0_1(y, x) = x$, $x(y, 0_2) = y(x, 0_2)$ and $x(0_3, x) = 0_3$. Then $0_1(0_3, 0_1) = 0_1$ and $0_1(0_3, 0_1) = 0_3$, hence $0_1 = 0_3$, say $= 0$. It follows that $0_2 = 0(0_2, 0_2) = 0_2(0, 0_2) = 0$.

Now let us consider the operation

(22) $(x, y)z = x \wedge y \rightarrow z = x' \vee y' \vee z$,

which satisfies $(x, x)y = x \rightarrow y$, hence

(23)
$$x \vee y = ((x, x)y, (x, x)y)y , \quad x' = (x, x)0 ,$$
$$x \wedge y = (((x, x)y, (x, x)y)y, ((x, x)y, (x, x)y))0 .$$

Nieminen [1976] constructed the system

$$\mathbf{B}_{63} = \{B_{111}, B_{112}, B_{113}, B_{114}\} ,$$

where

B_{111} $((x, x)y, (x, x)y)x = x$,

B_{112} $((x, x)y, (x, x)y))y = ((y, y)x, (y, y)x)x$,

B_{113} $(x, x)((y, y)z) = (y, y)((x, x)z)$,

B_{114} $(0, 0)x = (x, 0)0$.

In order to parry the non-associativity of the Sheffer stroke, Moisil [1959] introduced the *Sheffer functions of n variables*,

$$(24) \qquad \perp (x_1, \ldots, x_n) = x'_1 \wedge \ldots \wedge x'_n, \ (n = 1, 2, \ldots)$$

and their duals, which he used in the study of circuits involving electronic tubes, transistors or cryotrons. Answering Moisil's call to an algebraic study of Sheffer functions of n variables, Hammer [1959] defined Boolean algebras in terms of these functions. He devised the system

$$\mathbf{B}_{64} = \{ \mathrm{B}^{\perp}_{115}, \mathrm{B}^{\perp}_{116}, \mathrm{B}^{\perp}_{117}, \mathrm{B}^{\perp}_{118}, \mathrm{B}^{\perp}_{119}, \mathrm{B}^{\perp}_{120}, \mathrm{B}^{\perp}_{121} \},$$

where

$\mathrm{B}^{\perp}_{115}$ $\quad \perp (x, x) = \perp x$,

$\mathrm{B}^{\perp}_{116}$ $\quad \perp (\perp\perp (x_1, \ldots, x_{n-1}), x_n) = \perp (x_1, \ldots, x_n)$,

$\mathrm{B}^{\perp}_{117}$ $\quad \perp (x, y) = \perp (y, x)$,

$\mathrm{B}^{\perp}_{118}$ $\quad \perp (x, \perp (x, y)) = \perp (x, \perp y)$,

$\mathrm{B}^{\perp}_{119}$ $\quad \perp\perp x = x$,

$\mathrm{B}^{\perp}_{120}$ $\quad \perp (x, \perp x) = \perp (y, \perp y)$,

$\mathrm{B}^{\perp}_{121}$ $\quad \perp (\perp (x, y), \perp (x, z)) = \perp\perp (x, \perp (\perp (\perp y, \perp z)))$;

the Boolean algebra is obtained via

$$(25) \qquad x \vee y = \perp\perp (x, y), \ x \wedge y = \perp (\perp x, \perp y), \ x' = \perp x.$$

4.7. Boolean Algebras in Terms of Relations

The first attempts to define Boolean algebras using the relation of partial order $\leq$ go back to Peirce [1884] and Schröder [1890-1905], vol.I, but these authors didn't provide rigorous systems of postulates. Hahn [1909] devised another unsatisfactory Schröder-like system, and so did Pereira [1951]*, who partially improved Schröder's system. Church [1952] criticized the still unsatisfactory Schröder-Pereira system and suggested a system of his own, in terms of both the relation $\leq$ and the operations $\vee, \wedge, '$ together with 0 and 1.

As a matter of fact, it was Huntington [1904] who obtained the first definition of Boolean algebras based on partial order $\leq$, complementation $'$ and the elements 0,1, in his *second system*,

$$\mathbf{B}_{65} = \mathbf{II} = \{ \mathrm{B}_{122}, \mathrm{B}_{123}, \mathrm{B}_{124}, \mathrm{B}^0_{125}, \mathrm{B}^1_{125}, \mathrm{B}^{\vee}_{126}, \mathrm{B}^{\wedge}_{126}, \mathrm{B}^{\vee}_{127}, \mathrm{B}^{\wedge}_{127}, \mathrm{B}_{128} \},$$

where B_{122}, B_{123}, B_{124} are the axioms P_1, P_2, P_3 of partial order, B_{125} are the properties of 0 and 1, B_{126} state the existence of join and meet, and

$B_{127}^{\vee}$ $x \leq y \ \& \ x' \leq y \Longrightarrow y = 1$,

$B_{127}^{\wedge}$ $y \leq x \ \& \ y \leq x' \Longrightarrow y = 0$,

B_{128} $x \not\leq y' \Longrightarrow \exists z \neq 0 \ (z \leq x \ \& \ z \leq y)$.

Couturat [1914]* provided the following variant of system **II**:

$$\mathbf{B}_{66} = \{B_{123}, B_{124}, B_{125}^0, B_{125}^1, B_{126}^{\vee}, B_{126}^{\wedge}, B_{129}, B_{10}, B_{130}\} ,$$

in which he introduced the notation $x \vee y$ and $x \wedge y$ for the elements described in axioms B_{126} and

B_{129} $(x \vee y) \wedge z \leq (x \wedge z) \vee (y \wedge z)$,

B_{130} $1 \not\leq 0$.

The elements 0 and 1 can be dispensed with, as shown by MacNeille [1937] in his system

$$\mathbf{B}_{67} = \{B_{123}, B_{124}, B_{126}^{\wedge}, B_{131}^{\wedge}, B_{132}^{\wedge}, B_{133}\} ,$$

where

$B_{131}^{\wedge}$ $x \wedge x' \leq y$,

$B_{132}^{\wedge}$ $x \wedge y \leq z \ \forall z \Longrightarrow x \leq y'$,

B_{133} $x \leq y \Longleftrightarrow y' \leq x'$.

Another way of avoiding the elements 0 and 1 is to replace them by the unary relations $Nx \Longleftrightarrow \forall z \ x \leq z$ and $Ux \Longleftrightarrow \forall z \ z \leq x$. Byrne [1948] defined Boolean algebras in terms of relations N and U together with the binary relation

(1) $Cxy \Longleftrightarrow (z \leq x \ \& \ z \leq y \Longrightarrow Nz) \ \& \ (x \leq z \ \& \ y \leq z \Longrightarrow Uz)$,

which expresses the property $x \wedge y = 0 \ \& \ x \vee y = 1$, and the ternary relation $Pxyz$ which translates the fact that $x \wedge y = z$. He devised the system

$$\mathbf{B}_{68} = \{B_{123}, B_{124}, B_{134}, B_{135}, B_{136}\} ,$$

where

B_{134} $\forall x \forall y \exists p \exists p' \ Pxyp \ \& \ Cpp'$,

B_{135} $\exists z \ (Cyz \ \& \ Pxzp \ \& \ Np) \Longrightarrow x \leq y$,

B_{136} $\exists x \exists y \ x \not\leq y$.

However a simpler system had been given ten years earlier, in terms of the relations $\leq, N$ and

(2) $$xDy \Longleftrightarrow (z \le x \ \& \ z \le y \Longrightarrow Nz) \ ,$$

which is a translation of the *disjointness relation* $xDy \Longleftrightarrow x \wedge y = 0$.

Proposition 4.7.1. (Tarski [1938]) *Boolean algebras are characterized by the system*

$$\mathbf{B}_{69} = \{ \mathbf{B}_{123}, \mathbf{B}_{124}, \mathbf{B}_{126}^{\vee}, \mathbf{B}_{137} \} \ ,$$

where

$\mathbf{B}_{137} \qquad \forall x \, \exists y \ xDy \ \& \ (xDz \Longrightarrow z \le y) \ \& \ (zDy \Longrightarrow z \le x) \ .$

PROOF: We are checking Huntington's system **II**.

Taking $z := x$ in $\mathbf{B}_{137}$ it follows that $x \le x$ and so $\le$ is a partial order.

For a given element x, let y and y_1 be two elements satisfying $\mathbf{B}_{137}$. Then xDy_1 and $xDy_1 \Longrightarrow y_1 \le y$, hence $y_1 \le y$ and similarly $y \le y_1$, showing that $y = y_1$. So the element y in $\mathbf{B}_{137}$ is uniquely determined by x and we can set $y = x'$, which introduces a unary operation $'$. Therefore $\mathbf{B}_{137}$ can be paraphrased in the form

(3.1) $$xDx' \ ,$$

(3.2) $$xDz \Longrightarrow z \le x' \ ,$$

(3.3) $$zDx' \Longrightarrow z \le x \ .$$

Definition (2) shows that relation D is symmetric, hence from (3.1) we infer $x'Dx$, whence (3.2) implies $x \le x''$. Moreover, from $x''Dx'$ we deduce $x'' \le x$ by (3.3). Therefore $x'' = x$.

Now we need to prove that $x \le y \Longrightarrow y' \le x'$. For suppose $x \le y$. Suppose z satisfies $z \le x$ and $z \le y'$. Then $z \le y$ and since yDy' it follows by (2) that Nz; this proves that xDy', again by (2), therefore $y' \le x'$ by (3.2).

From now on we are going to use freely $x'' = x$ and $x \le y \Longrightarrow y' \le x'$.

First we denote by $x \vee y$ the least upper bound of x and y, whose existence is stated in $\mathbf{B}_{126}^{\vee}$. Then we set $x \wedge y = (x' \vee y')'$. This implies $x' \wedge y' = (x \vee y)'$, hence $x \vee y = (x' \wedge y')'$. Besides, $x' \le x' \vee y' = (x \wedge y)'$, hence $x \wedge y \le x$ and similarly $x \wedge y \le y$. Moreover, if $z \le x$ and $z \le y$, then from $x' \le z'$ and $y' \le z'$ we infer $x' \vee y' \le z'$, hence $z \le (x' \vee y')' = x \wedge y$. We have thus proved $\mathbf{B}_{126}^{\wedge}$, where the greatest lower bound of x and y is $x \wedge y$.

In particular $x \wedge x' \le x$ and $x \wedge x' \le x'$. But xDx', therefore $N(x \wedge x')$ by (2). So $x \wedge x'$ is the least element, say $= 0$. From $x \wedge x' = 0$ we deduce

$x \vee x' = (x' \wedge x'')' = 0'$. Setting $0' = 1$, we have $x \vee x' = 1$. Since $0 \leq x'$ for all x, it follows that $x \leq 0'$, that is, 1 is the greatest element.

Summarizing, we have proved properties B_{125}. Besides, definition (2) reads $xDy \Longleftrightarrow x \wedge y = 0$. To prove B_{128}, we write it in the form $x \nleq y' \Longrightarrow x \wedge y \neq 0$. This implication follows from the fact that if $x \wedge y = 0$ then xDy, hence yDx, therefore $x \leq y'$ by (3.2) $\square$

Going back in time, we find a paper by Yule [1926], in which Boolean algebras are defined solely in terms of the disjointness relation D. This is possible because in every Boolean algebra

$$x \leq y \Longrightarrow \forall z \, (y \wedge z = 0 \Longrightarrow x \wedge z = 0)$$

$$\Longrightarrow \forall z \, (z \leq y' \Longrightarrow z \leq x') \Longrightarrow y' \leq x' \Longrightarrow x \leq y \,.$$

The system provided by Yule is

$$\mathbf{B}_{70} = \{B_{138}, B_{139}, B_{140}, B_{141}\} \,,$$

where

$B_{138} \qquad \forall z \, (yDz \Longrightarrow xDz) \, \& \, \forall z \, (xDz \Longrightarrow yDz) \Longrightarrow x = y \,,$

$B_{139} \qquad xDx \Longrightarrow \forall z \, xDz \,,$

$B_{140} \qquad \forall x \, \exists x' \; xDx' \, \& \, (xDy \Longrightarrow \forall z \, (x'Dz \Longrightarrow yDz)) \,,$

$B_{141} \qquad \forall x \, \forall y \, \exists p \qquad \forall z \, (pDz \Longrightarrow qDz) \Longleftrightarrow$

$\qquad\qquad \Longleftrightarrow \forall z \, (xDz \Longrightarrow qDz) \, \& \, \forall z \, (yDz \Longrightarrow qDz) \,.$

The order relation is defined by

(4) $\qquad\qquad\qquad\qquad x \leq y \Longleftrightarrow \forall z \, (yDz \Longrightarrow xDz) \,,$

which makes the meaning of the axioms quite clear.

4.8. Huntington Varieties

It has been known for a long time that in a bounded distributive lattice, an element can have at most one complement; cf. Proposition 4.1.1. In particular, every element of a Boolean algebra is uniquely complemented. In 1904 Huntington conjectured the converse, that any uniquely complemented lattice was distributive. In fact, the conjecture had been verified for several special classes of lattice. However, in 1945 Dilworth disproved this conjecture by proving that any lattice can be embedded into a uniquely complemented lattice. In 1969, Chen and Grätzer showed that this particular result can be obtained without making use of some of the more difficult machinery developed in the above paper (e.g., an extra unary operator). In 1981, Adams and Sichler strenghtened the original embedding theorem of Dilworth by showing the existence of a continuum of varieties in which each lattice can be embedded in a uniquely complemented lattice of the same variety. In spite of these deep theorems, it is still hard to find "nice" and "natural" examples of uniquely complemented lattices that are not Boolean. The reason is that uniquely complemented lattices having a little extra structure most often turn out to be distributive. This seems to be the essence of Huntington's conjecture. Accordingly, we plan to attack the problem backwards: that is, by finding additional (albeit, mild) condition that, if added, would solve the problem in the affirmative. Many such conditions were discovered during 1930s and 1940s. The most notable among such conditions, due to Birkhoff and von Neumann, is modularity. Let us call a lattice property P, a *Huntington property* if every uniquely complemented P-lattice is distributive. Similarly, a lattice variety K is said to be a *Huntington variety* if every uniquely complemented member of K is distributive. The monograph by Salii [1988] compiles a number of known Huntington properties. Theorem 8.1 below gives some of the important Huntington properties that we will employ in this section to get non-trivial axiomatizations of Boolean algebras as uniquely complemented lattices.

Remark 4.8.1. In every uniquely complemented lattice identity $x'' = x$ holds, because both x and x'' are complements of x'. Therefore condition 1 in Theorem 8.1 below is equivalent to $x \leq y \iff y' \leq x'$, while the two De Morgan laws are equivalent. The following *Boolean absorption laws* are also equivalent under the De Morgan laws:

BA$^\vee$ $(x \wedge y') \vee y = x \vee y$,

BA$^\wedge$ $(x \vee y') \wedge y = x \wedge y$.

Theorem 4.8.1. *If L is a uniquely complemented lattice, then the following statements are equivalent:*

1. *L is modular;*
2. *L has order reversible complementation, i.e., $x \leq y \implies y' \leq x'$;*
3. *L satisfies one of the De Morgan laws;*
4. *L satisfies one of the Boolean absorption laws;*
5. *L satisfies the Frink implication $x' \vee y = 1 \implies x \leq y$;*
6. *L is Boolean.*

PROOF: We will use freely Remark 8.1.

$1 \implies 2$: Assume $a \leq b$. Modularity implies $b = b \wedge (a' \vee a) = (b \wedge a') \vee a$, hence

$$a \vee ((a' \wedge b) \vee b') = a \vee (a' \wedge b) \vee ((a' \wedge b) \vee a)' = 1 \; ;$$

by applying once again modularity we obtain

$$a \wedge ((a' \wedge b) \vee b') = a \wedge b \wedge (b' \vee (a' \wedge b)) = a \wedge ((b \wedge b') \vee (a' \wedge b)) = 0 \; .$$

Since the complement of a is unique, it follows from the above two identities that $a' = (a' \wedge b) \vee b' \geq b'$.

$2 \Longrightarrow 3^{\dagger}$: From $x \wedge y \leq x, y$ we infer $x', y' \leq (x \wedge y)'$, hence $x' \vee y' \leq (x \wedge y)'$. Similarly, from $x', y' \leq x' \vee y'$ we infer $(x' \vee y')' \leq x \wedge y$, hence $(x \wedge y)' \leq x' \vee y'$, therefore $(x \wedge y)' = x' \vee y'$.

$3 \Longrightarrow 5$: Suppose $a' \vee b = 1$ and set $X = (a' \wedge b) \vee (b \wedge (a \vee b'))$. Then

$$b' \vee X = b' \vee (a' \wedge b) \vee (b' \vee (a' \wedge b))' = 1 \; ,$$

while $X \leq b$, hence $b' \wedge X \leq b' \wedge b = 0$. So $b' \vee X = 1$ and $b' \wedge X = 0$, whence $X = b$ by the uniqueness of the complement. It follows that $a' \vee X = a' \vee b = 1$, that is, $a' \vee (b \wedge (a \vee b')) = 1$. Since we have also

$$a' \wedge (b \wedge (a \vee b')) = a' \wedge b \wedge (a' \wedge b)' = 0 \; ,$$

the uniqueness of the complement implies $b \wedge (a \vee b') = a$. Taking the meet of each side with a we obtain $a = a \wedge b \wedge (a \vee b') = a \wedge b \leq b$.

$5 \Longrightarrow 6$: Since the converse of the Frink implication holds trivially, the hypothesis 5 reads $x' \vee y = 1 \Longleftrightarrow x \leq y$, Hence L is a Boolean algebra by Theorem 3.1.

$6 \Longrightarrow 1$ and $6 \Longrightarrow 4$: Trivial.

$4 \Longrightarrow 2$: If $x \leq y$ then by taking the join of each side with x' we obtain $x' \vee y = 1$, whence $\mathrm{BA}^{\wedge}$ yields $x' \wedge y' = (x' \vee y) \wedge y' = y'$, hence $y' \leq x'$. $\qquad \square$

The following comments are in order. The equivalence $1 \Longleftrightarrow 6$ is due to Birkhoff and von Neumann; see Salii [1988], page 40. Characterization 3 generalizes Theorem X.17 in Birkhoff [1948]. Properties 2-5 in Theorem 8.1 are in fact among the most useful computational properties of Boolean algebras, just like B_1-B_9. From another point of view, properties 2 and 5 are (Boolean) equational implications, properties 3 and 4 are Boolean identities, while modularity is the only lattice identity in the list. There are also several Huntington properties of a different nature, for instance finiteness, being atomic, having a finite order-dimension, or semimodularity.

---

†This implication does not require uniqueness of the complement and is also valid in orthocomplemented lattices; cf. Salii [1988].

But so far modularity was the only known lattice identity (in terms of $\vee$ and $\wedge$ only) having the Huntington property.

In this section we give several new lattice identities and implications that are Huntington properties and finally we generalize the classical result 5 of Birkhoff and von Neumann by providing many lattice identities which force a uniquely complemented lattice to be Boolean.

Theorem 4.8.2. *The variety generated by* N5 *is a Huntington variety.*

PROOF: A specialization of the four-variable identity (η_2) in McKenzie [1972], p.7, forming part of an equational basis for the variety N5, yields the following identity:

(1) $$x \wedge (y \vee z) = (x \wedge (y \vee (x \wedge z))) \vee (x \wedge (z \vee (x \wedge y))) .$$

Now let $a \wedge b' = 0$. Then (1) yields

$$b' \wedge (b \vee a) = (b' \wedge (b \vee (b' \wedge a))) \vee (b' \wedge (a \vee (b' \wedge b))) = (b' \wedge b) \vee (b' \wedge a) = 0$$

and, of course, $b' \vee (b \vee a) = 1$. So $b \vee a$ and b are complements of b', therefore $b \vee a = b$. We have thus proved that $a \wedge b' = 0 \implies a \leq b$. Conversely, $a \leq b$ implies $a \wedge b' \leq b \wedge b' = 0$, that is, $a \wedge b' = 0$. So property 2 in Theorem 8.1 is fulfilled, therefore the lattice is a Boolean algebra. □

The lattice N5 is the simplest example of what is known as a splitting lattice. It is a finite sulattice of a free lattice and it is subdirectly irreducible. More generally, we can show that any variety of lattices generated by a splitting lattice is Huntington: i.e., if a uniquely complemented lattice belongs to a variety generated by a splitting lattice then it must be a Boolean algebra. What really happens here is that such varieties do satisfy a lattice equation (like (1) above) which formally implies $\mathrm{SD}^\vee$ or $\mathrm{SD}^\wedge$ below which, in turn, force distributivity under unique complementation. Thus there are infinitely many lattice varieties which are Huntington in this sense; for more details see Salii [1988] and Padmanabhan, McCune and Veroff [2007].

Theorem 4.8.3. *A uniquely complemented lattice satisfying any of the two implications below is a Boolean algebra:*

$\mathrm{SD}^\wedge$ $\qquad x \wedge y = x \wedge z \implies x \wedge y = x \wedge (y \vee z)$;

$\mathrm{CM}^\vee$ $\qquad x \vee y = x \vee z \implies x \wedge ((x \wedge y) \vee z) = (x \wedge y) \vee (x \wedge z)$.

PROOF: We are going to prove order reversibility of $'$, which implies distributivity by Theorem 8.1.1. Let $a \leq b$. Since $a \vee (a' \vee b') = 1$, if we succeed to prove that $a \wedge (a' \vee b') = 0$, the uniqueness of the complement of a will imply $a' \vee b' = a'$, that is $b' \leq a'$, as desired.

But $a \leq b$ implies $a \wedge b' \leq b \wedge b' = 0$, hence $a \wedge a' = 0 = a \wedge b'$. If SD$^\wedge$ holds, this implies $0 = a \wedge a' = a \wedge (a' \vee b')$. Now suppose CM$^\vee$ holds. From $1 = a' \vee a \leq a' \vee b$ we get $b \vee a' = 1 = b \vee b'$, hence

$$b \wedge ((b \wedge a') \vee b') = (b \wedge a') \vee (b \wedge b') = b \wedge a'$$

and taking the meet with a on each side we obtain

$$(2) \qquad\qquad a \wedge ((b \wedge a') \vee b') = 0 .$$

On the other hand $b \vee a' = 1 = b \vee (a \vee b')$, hence CM$^\vee$ implies

$$(3) \qquad\qquad b \wedge ((b \wedge a') \vee (a \vee b')) = (b \wedge a') \vee (b \wedge (a \vee b'))$$

and interchanging the roles of a' and $a \vee b'$ we obtain

$$(4) \qquad\qquad b \wedge ((b \wedge (a \vee b')) \vee a') = (b \wedge (a \vee b')) \vee (b \wedge a') .$$

Using lattice absorption L$_4^\wedge$, $a \leq b$, (4) and (3), we infer

$$(5) \qquad \begin{aligned} b &= b \wedge (a \vee a') = b \wedge ((a \wedge (a \vee b')) \vee a') \leq b \wedge ((b \wedge (a \vee b')) \vee a') \\ &= (b \wedge (a \vee b')) \vee (b \wedge a') = b \wedge ((b \wedge a') \vee (a \vee b')) \leq b . \end{aligned}$$

So all the elements occurring in (5) are equal to b. In particular $b = b \wedge ((b \wedge a') \vee a \vee b')$, that is, $b \leq (b \wedge a') \vee a \vee b'$. Therefore $1 = b \vee b' \leq (b \wedge a') \vee a \vee b'$, that is, $a \vee ((b \wedge a') \vee b') = 1$. From this identity and (2) we infer that $(b \wedge a') \vee b'$ is a complement of a, therefore $(b \wedge a') \vee b' = a'$, hence $b' \leq a'$.   □

Theorem 4.8.4. *A uniquely complemented lattice satisfying any of the the identities*

$$(6) \qquad x \wedge ((x \wedge y) \vee ((x \wedge z) \vee (y \wedge (x \vee z)))) = (x \wedge y) \vee (x \wedge z) ,$$

$$(7) \qquad x \wedge ((x \wedge y) \vee (z \wedge (x \vee (y \wedge (x \vee z))))) = (x \wedge y) \vee (x \wedge z) ,$$

$$(8) \qquad x \wedge ((y \wedge (x \vee z)) \vee (z \wedge (x \vee y))) = (x \wedge y) \vee (x \wedge z) ,$$

is a Boolean algebra.

PROOF: We check condition CM$^\vee$ in Theorem 8.3. Suppose $x \vee y = x \vee z$. If (6) holds then

$$(x \wedge z) \vee (x \wedge y) = x \wedge ((x \wedge z) \vee ((x \wedge y) \vee (z \wedge (x \vee z))))$$

$$= x \wedge ((x \wedge z) \vee ((x \wedge y) \vee z)) = x \wedge (z \vee (x \wedge y)) \ .$$

If (7) holds then

$$(x \wedge y) \vee (x \wedge z) = x \wedge ((x \wedge y) \vee (z \wedge (x \vee (y \wedge (x \vee y)))))$$

$$= x \wedge ((x \wedge y) \vee (z \wedge (x \vee y))) = x \wedge ((x \wedge y) \vee (z \wedge (x \vee z))) = x \wedge ((x \wedge y) \vee z) \ .$$

For (8) see Appendix A. □

Finally we have the promised generalization of the theorem on modularity:

Corollary 4.8.1. *A uniquely complemented lattice belonging to the variety* M $\vee$ N5 *is a Boolean algebra.*

PROOF: Every modular lattice satisfies the Huntington identity (7) because

$$x \wedge ((x \wedge y) \vee (z \wedge (x \vee (y \wedge (x \vee z))))) = x \wedge ((x \wedge y) \vee (z \wedge ((x \vee y) \wedge (x \vee z))))$$

$$= x \wedge ((x \wedge y) \vee (z \wedge (x \vee y))) = x \wedge (((x \wedge y) \vee z) \wedge (x \vee y))$$

$$= x \wedge ((x \wedge y) \vee z) \wedge (x \vee y) = ((x \wedge y) \vee z) \wedge x = (x \wedge y) \vee (z \wedge x) \ .$$

The lattice N5 satisfies (7) as well. This can be checked by considering the following 12 cases: $x, y, z \in \{0, 1\}$; $x = y$, $x = z$, $y = z$; x, y, z in the role of the element having 2 complements.

Therefore by Theorem 8.4 the variety M$\vee$N5 is Huntington. □

Finally we use the Huntington property of order reversibility in order to give one more example of a Huntington identity.

Theorem 4.8.5. *The non-modular identity*

$$(9) \qquad x \wedge (y \vee (x \wedge ((y \wedge x) \vee (z \wedge (y \vee x))))) = x \wedge (y \vee (x \wedge z))$$

defines a Huntington variety.

PROOF: In view of Theorem 8.1 it suffices to prove that complementation is order-reversing. So assume $a \leq b$. Note first that $a' \vee b = 1$. Hence taking $x := a', y := b$ in (9) yields

$$a' \wedge (b \vee (a' \wedge ((b \wedge a') \vee z))) = a' \wedge (b \vee (a' \wedge z)) \ ,$$

which for $z := (b \wedge a')'$ implies further

$$a' = a' \wedge (b \vee (a' \wedge (b \wedge a')')) \leq b \vee (a' \wedge (b \wedge a')') \ ,$$

whence by taking join of both sides with b we obtain $b \vee (a' \wedge (b \wedge a')') = 1$. Since on the other hand $b \wedge (a' \wedge (b \wedge a')') = 0$, it follows that $a' \wedge (b \wedge a')' = b'$ by the uniqueness of the complement. So $b' \leq a'$, as desired.

The above identity is non-modular because it is valid in N5. $\square$

4.9. Boolean Algebras Are Always One-Based

In the previous sections we saw various treatments of Boolean algebras and in most cases we could define the equational theory by a single axiom. This is no accident: the equational theory of Boolean algebras has a single axiom, whatever be the type. This follos from a more general theorem in universal algebra, as we are going to show in this section. Firstly we recall several prerequisites, following Grätzer [1978], Theorems I.3.9, II.3.11 and III.3.9.

A *congruence* relation of an algebra (A, F) is an equivalence relation on A which is compatible with all the operations in F. The set $\mathrm{Con}(A)$ of all the congruences on A is made into a lattice $(\mathrm{Con}(A), \cap, \vee)$, where $\cap$ is the set-theoretical intersection, that is, $x (\theta \cap \varphi) y \iff x \theta y \, \& \, x \varphi y$, and $\theta \vee \varphi$ is the intersection of all congruences which include θ and φ. The lattice $\mathrm{Con}(A)$ is called the *congruence lattice* of A. Now consider the case when the algebra is a lattice.

Lemma 4.9.1. *If σ is a congruence of a lattice, then*

(1) $$x \leq y \leq z \, \& \, x \sigma z \Longrightarrow x \sigma y \sigma z \, ;$$

(2) $$(x \wedge y) \sigma (x \vee y) \iff x \sigma y \Longrightarrow x \sigma (x \wedge y) \, \& \, x \sigma (x \vee y) \, .$$

PROOF: (1): We have $x = x \wedge y \sigma z \wedge y = y$, hence $y \sigma x \sigma z$.

(2): If $x \sigma y$, then $x = x \wedge x \sigma (x \wedge y)$ and similarly $x \sigma (x \vee y)$. Therefore

$$x \sigma y \Longrightarrow x \sigma (x \wedge y) \, \& \, x \sigma (x \vee y) \Longrightarrow (x \wedge y) \sigma (x \vee y)$$

and it remains to prove the converse of the last implication. But $(x \wedge y) \sigma (x \vee y)$ implies $(x \vee (x \wedge y)) \sigma (x \vee (x \vee y))$, that is, $x \sigma (x \vee y)$. By interchanging x and y we obtain $y \sigma (y \vee x)$, therefore $x \sigma y$. $\square$

Proposition 4.9.1. *For any lattice L, the operation $\vee$ of the congruence lattice $\mathrm{Con}(L)$ can be described as follows: $x(\theta \vee \varphi)y$ if and only if there is a sequence $z_0, z_1, \ldots, z_n$ such that*

(3)
$$z_0 = x \wedge y \leq z_1 \leq \cdots \leq z_i \leq z_{i+1} \leq \cdots \leq z_n = x \vee y$$
$$\text{and } z_i \, (\theta \cup \varphi) \, z_{i+1} \, (i = 0, 1, \ldots, n-1) \, ,$$

where $a \, (\theta \cup \varphi) \, b$ *means* $a \, \theta \, b$ *or* $a \, \varphi \, b$.

PROOF: Routine. See e.g. Grätzer, Theorem I.3.9. □

Theorem 4.9.1. (Funayama-Nakayama) *The congruence lattice of every lattice is distributive.*

PROOF: Let $\sigma, \theta, \varphi \in \mathrm{Con}(L)$. It suffices to prove that $\sigma \cap (\theta \vee \varphi) \subseteq (\sigma \cap \theta) \vee (\sigma \cap \varphi)$, since the converse inclusion holds in any lattice. Suppose $x \, (\sigma \cap (\theta \vee \varphi)) \, y$. Then $x \, \sigma \, y$ and let $z_0, z_1, \ldots, z_n$ be a sequence satisfying (3). But $(x \wedge y) \, \sigma \, (x \vee y)$ by Lemma 9.1(2), that is, $z_0 \, \sigma \, z_n$, therefore $z_i \, \sigma \, z_{i+1} \, (i = 0, \ldots, n-1)$ by Lemma 9.1(1). But using also (3), we infer that for each i we have $z_i \, (\sigma \cap \theta) \, z_{i+1}$ or $z_i \, (\sigma \cap \varphi) \, z_{i+1}$, that is, $z_i \, ((\sigma \cap \theta) \cup (\sigma \cap \varphi)) \, z_{i+1}$. Therefore $x \, ((\sigma \cap \theta) \cup (\sigma \cap \varphi)) \, y$. □

Recall that by an *ideal* of a lattice L is meant a non-mpty subset I of L such that $x, y \in I \implies x \vee y \in I$ and $x \leq y \in I \implies x \in I$. If L is a lattice with least element 0, it is easily seen that for every congruence σ of L, the set $id(\sigma) = \{x \in L \mid x \, \sigma \, 0\}$ is an ideal of L. It is also easy to see that for every ideal I of a distributive lattice L, the relation $cn(I)$ defined by $x \, cn(I) \, y \iff \exists i \in I \, x \vee i = y \vee i$, is a congruence of L. These properties are much strengthened in the case of a Boolean algebra.

As prescribed by universal algebra, the congruences of a Boolean algebra are those congruences σ of the lattice reduct $(B, \vee, \wedge)$ which are also compatible with complementation, that is, $x \, \sigma \, y \implies x' \, \sigma \, y'$. However we will see that Boolean congruences coincide with the congruences of the lattice reduct. The ideals of the Boolean algebra are defined as being the ideals of the lattice reduct.

In the sequel we need the Boolean identity

(4)
$$x \vee (x + y) = x \vee y = y \vee (x + y) \, ,$$

which follows by (4.1), $\mathrm{B}_4^\vee$ and Boolean absorption.

Lemma 4.9.2. *Let I be an ideal of a Boolean algebra B. Then*

$$x \, cn(I) \, y \iff x + y \in I \, .$$

PROOF: If $x \, cn(I) \, y$ then $x' \wedge i' = y' \wedge i'$ for some $i \in I$, hence $x \wedge y' \wedge i' = 0$, hence $x \wedge y' \leq i \in I$, implying $x \wedge y' \in I$, and similarly $x' \wedge y \in I$, therefore

$x + y \in I$ by (4.1). Conversely, if $x + y \in I$ then it follows from (4) that $x \, cn(I) \, y$. $\qquad \square$

Corollary 4.9.1. $cn(I)$ *is a congruence of the Boolean ring* B.

Proposition 4.9.2. *In every Boolean algebra the mappings* $\sigma \mapsto id(\sigma)$ *and* $I \mapsto cn(I)$ *establish a bijection between congruences and ideals.*

PROOF: We already know that $id(\sigma)$ is an ideal and $cn(I)$ is a lattice congruence. The properties $cn(id(\sigma)) = \sigma$ and $id(\sigma(I)) = I$ are easily checked using (4), Lemma 9.2 and Corollary 9.1. For instance, let us check the former property. If $x \, \sigma \, y$ then $(x \wedge y') \, \sigma \, (y \wedge y') = 0$ and $(x' \wedge y) \, \sigma \, 0$, hence $(x + y) \, \sigma \, (0 \vee 0) = 0$, that is, $x + y \in id(\sigma)$, therefore (4) implies $x \, cn(id(\sigma)) \, y$. Conversely, the latter relation implies $x \vee i = y \vee i$ for some $i \in id(\sigma)$, that is, $i \, \sigma \, 0$. So, denoting by $[a]$ the equivalence class of an element a modulo σ, we have $[i] = [0]$ and

$$[x] = [x \vee 0] = [x] \vee [0] = [x] \vee [i] = [x \vee i] = [y \vee i] = \cdots = [y] \, ,$$

that is, $x \, \sigma \, y$. $\qquad \square$

Proposition 4.9.3. *The congruences of every Boolean algebra* $(B, \vee, \wedge, ', 0, 1)$ *coincide with the congruences of the lattice reduct* $(B, \vee, \wedge)$ *of* B.

PROOF: In view of Proposition 9.2, it remains to prove that the congruences $cn(I)$ are Boolean. This follows from Lemma 9.2: if $x \, cn(I) \, y$ then $x' + y' = x + 1 + y + 1 = x + y \in I$, therefore $x' \, cn(I) \, y'$. $\qquad \square$

Corollary 4.9.2. *The congruence lattice of every Boolean algebra is distributive.*

PROOF: By Theorem 9.1 and Proposition 9.3 $\qquad \square$

Recall that the composition $\circ$ of binary relations is defined by

$$x \, (\theta \circ \varphi) \, y \iff \exists z \; x \, \theta \, z \; \& \; z \, \varphi \, y \, ,$$

and one says that θ and φ *permute*, or that they are *permutable*, provided $\theta \circ \varphi = \varphi \circ \theta$.

Lemma 4.9.3. *Let:* B *be a Boolean algebra,* θ *and* φ *congruences of* B, *and* $x, y, z \in B$ *such that* $x \leq y \leq z$, $x \, \theta \, y$ *and* $y \, \varphi \, z$. *Then there is* $u \in B$ *such that* $x \leq u \leq z$, $x \, \varphi \, u$ *and* $u \, \theta \, z$.

PROOF: Set $I = id(\theta)$ and $J = id(\varphi)$. Set also

$$j = x' \wedge y = (x' \wedge y) \vee (x \wedge y') = x + y$$

and note that $j \in I$ by Corollary 9.1, while (4) implies $x \vee j = x \vee y = y$. Similarly, setting $k = y' \wedge z$ we get $k \in J$ and $y \vee k = z$. Now take $u = x \vee k$. Then $x \leq u \leq x \vee z = z$ and $u \vee k = x \vee k$, hence $x \, \varphi \, u$. Besides, $j \leq y \leq z$, hence

$$z \vee j = z = y \vee k = x \vee j \vee k = u \vee j \, ,$$

therefore $u \, \theta \, z$. □

Theorem 4.9.2. *Every two congruences of a Boolean algebra permute.*

PROOF: Let $\theta, \varphi \in \mathrm{Con}(B)$. Since θ and φ can be interchanged, it suffices to prove that $a \, (\theta \circ \varphi) \, b \Longrightarrow a \, (\varphi \circ \theta) \, b$. Suppose $a \, (\theta \circ \varphi) \, b$. Then $a \, \theta \, c$ and $c \, \varphi \, b$ for some $c \in B$. We are going to find $d \in B$ such that $a \, \varphi \, d$ and $d \, \theta \, b$.

Since $a \, \theta \, (a \vee c)$ and $c \, \varphi \, (b \vee c)$, whence $(a \vee c) \, \varphi \, (a \vee b \vee c)$, we can apply Lemma 9.3 and find an element e such that $a \leq e \leq a \vee b \vee c$, $a \, \varphi \, e$ and $e \, \theta \, (a \vee b \vee c)$.

Since $b \, \varphi \, (b \vee c)$ and $b \, \theta \, (a \vee b)$, we apply again Lemma 9.3 and obtain an element f such that $b \leq f \leq a \vee b \vee c$, $b \, \theta \, f$ and $f \, \varphi \, (a \vee b \vee c)$.

Now take $d = e \wedge f$. But $(e \wedge f) \, \varphi \, (e \wedge (a \vee b \vee c))$, that is, $d \, \varphi \, e$, and since $a \, \varphi \, e$, it follows that $a \, \varphi \, d$. Also, $(f \wedge e) \, \theta \, (f \wedge (a \vee b \vee c))$, that is, $d \, \theta \, f$, and since $b \, \theta \, f$, it follows that $d \, \theta \, b$. □

Remark 4.9.1. A much shorter proof of Theorem 9.2 follows from Corollary 9.1 and the fact that the congruences of every group permute.

If the lattice $\mathrm{Con}(A)$ of an algebra A is distributive, then one says that A is *congruence distributive*, and if every two congruences of A permute, then A is said to be *congruence permutable*. If all the members of a variety have a certain common property, then one says that the variety itself has that property. In particular by an *arithmetical variety* is meant a variety which is both congruence distributive and congruence permutable.

Now from Theorems 9.1 and 9.2 we obtain

Corollary 4.9.3. *The variety of Boolean algebras is arithmetical.*

Pixley [1963] characterized congruence distributive and congruence permutable varieties by two ternary terms. Gould and Grätzer [1967] combined them into a single ternary term characterizing arithmetical varieties. Then Pixley [1971] independently rediscovered the latter result. This is

Theorem II.12.5 in Burris and Sankappanavar [1981], which says that a variety is arithmetical if and only if it admits a ternary polynomial p satisfying the following identities:

GGP $p(x, y, y) = x,\ p(x, y, x) = x,\ p(x, x, z) = z$,

to the effect that identities GGP hold in every algebra of that variety. In view of this, we refer to the above identity as GGP and to a ternary term satisfying it as a *Gould-Grätzer-Pixley term* or GGP term for short. It is important to note that being a GGP term is not an intrinsic property, but a property relative to a variety.

Proposition 4.9.4. *The unique* GGP *term of a Boolean algebra is*

$$(5) \qquad\qquad p(x, y, z) = (x \wedge y') \vee (y' \wedge z) \vee (x \wedge z) .$$

PROOF: In view of Löwenheim's Verification Theorem (see e.g. Rudeanu [1974], Theorem 2.13) an identity holds in a Boolean algebra if and only if it is verified for all the values 0,1 given to the variables. Hence identities GGP are equivalent to the following conditions:

$$(6.1) \qquad\qquad p(0,0,0) = p(0,1,1) = 0,\ p(1,0,0) = p(1,1,1) = 1 ,$$

$$(6.2) \qquad\qquad p(0,0,0) = p(0,1,0) = 0,\ p(1,0,1) = p(1,1,1) = 1 ,$$

$$(6.3) \qquad\qquad p(0,0,0) = p(1,1,0) = 0,\ p(0,0,1) = p(1,1,1) = 1 .$$

On the other hand, every Boolean polynomial satisfies the interpolation formula, which for 3 variables reads

$$(7) \qquad\qquad p(x, y, z) = \bigvee \{ p(\alpha, \beta, \gamma) \wedge x^{\alpha} \wedge y^{\beta} \wedge z^{\gamma} \mid \alpha, \beta, \gamma \in \{0,1\} \} ,$$

where we have set $x^0 = x'$ and $x^1 = x$; (see e.g. Rudeanu (op.cit), Theorem 1.6').

We see that conditions (6) are consistent. Therefore, in view of (7), they yield the unique GGP term

$$p(x, y, z) = (x \wedge y \wedge z) \vee (x \wedge y' \wedge z) \vee (x \wedge y' \wedge z') \vee (x' \wedge y' \wedge z) ,$$

which, after obvious simplifications which use the idempotency of $\vee$, reduces to (5).   □

Note that Proposition 9.4 provides a very short proof of Corollary 9.3.

The relevance of the above results to the axiomatics of Boolean algebras follows from Theorem 9.3 below, known since 1967 and due to Baker, Tarski,

McKenzie, Grätzer, Quackenbush and Padmanabhan. We follow here the proof given by Padmanabhan and Quackenbush [1973].

Theorem 4.9.3. *Every finitely based arithmetical variety is one-based.*

PROOF: We are going to use the GGP characterization of arithmetical varieties.

First we remark that any identity $f = g$ is equivalent to $p(u, f, g) = u$, where u is a variable not occurring in f or g. For $p(u, f, f) = u$ and conversely, if $p(u, f, g) = u$ holds, then we take $u := f$ and obtain $p(f, f, g) = f$; but $p(f, f, g) = g$ by (4), hence $f = g$.

Now let us prove that every two identities can be combined into a single identity. First we use the remark above and express the given identities as absorption identities, say $f(y, y_1, \ldots, y_n) = y$ and $g(z, z_1, \ldots, z_m) = z$, and we claim that the system $f = y$, $g = z$ is equivalent to

(8) $$p(x, f, y) = p(x, g, z) \ .$$

For the system implies

$$p(x, f, y) = p(x, y, y) = x = p(x, z, z) = p(x, g, z) \ .$$

Conversely, (8) implies $y = p(f, f, y) = p(f, g, f) = f$ and $g = p(g, f, f) = p(g, g, z) = z$.

A repeated application of the above transformations reduces any finite system of identities to a single absorption identity, say $f(z, z_1, \ldots, z_n) = z$. It remains to prove that the system consisting of the three axioms GGP and $f = z$ is equivalent to the single identity

(9) $$p(p(u, u, x), p(f, y, z), z) = x \ .$$

Clearly the system implies

$$p(p(u, u, x), p(f, y, z), z) = p(x, p(z, y, z), z) = p(x, z, z) = x \ .$$

Conversely, suppose (9) holds. Define the function $f_1(x, z_1, \ldots, z_n) = f(x, z_1, \ldots, z_n)$. Then (9) implies

$$p(p(f_1, f_1, x), p(f_1, x, x), x) = x \ ,$$

which shows that x can be written in the form $x = p(u, u, x)$, therefore (9) implies

(10) $$p(x, p(f, y, z), z) = x \ ,$$

whence

(11) $$p(p(u, u, x), p(f, y, z), z) = p(u, u, x) .$$

It follows from (9) and (11) that

(12) $$p(u, u, x) = x .$$

By applying in turn (12) and (10) we obtain

(13) $$z = p(p(f, y, z), p(f, y, z), z) = p(f, y, z) ,$$

therefore (10) becomes

(14) $$p(x, z, z) = x .$$

By applying in turn (13) and (14) we get

$$z = p(f, z, z) = f$$

so that (13) becomes $z = p(z, y, z)$, completing the proof of GGP.   □

Corollary 4.9.4. *Every finitely based variety which admits a GGP-term is one-based.*

PROOF: By Theorem 9.3 and Pixley's theorem.   □

Theorem 4.9.4. *The variety of Boolean algebras is one-based, whatever be the type.*

PROOF: Under term equivalence (i.e. definitional equivalence), the polynomials remain the same, and hence it is easy to see that the congruences are the same, so that the variety remains arithmetical. Since the finite basis property remains invariant under term-equivalence, the variety of all Boolean algebras – of any finite type – is obviously finitely based and hence is one-based by Theorem 9.3 and Corollary 9.3.   □

4.10. Orthomodular Lattices

Orthomodular lattices generalize Boolean algebras. They have arisen in the study of quantum logic, that is, the logic which supports quantum mechanics and which does not conform to classical logic. As noted by Birkhoff and von Neumann [1936], the calculus of propositions in quantum logic "is formally indistinguishable from the calculus of linear subspaces [of a Hilbert space] with respect to set products, linear sums and orthogonal complements" in the roles of *and, or* and *not*, respectively. This has led to the study of the closed subspaces of a Hilbert space, which form an orthomodular lattice in contemporary terminology. As oftenhappens in algebraic logic, the study of orthomodular lattices has tremendously developed, both for their interest in logic and for their own sake; see Kalmbach [1983].

This section is devoted to the axiomatics of orthonormal lattices and of the more general class of ortholattices. After the necessary definitions, we present a few systems of axioms and several metatheorems.

An *ortholattice* is an algebra $(L, \vee, \wedge, \,', 0, 1)$ such that the reduct $(L, \vee, \wedge, 0, 1)$ is a bounded lattice and the unary operation $'$ satisfies the De Morgan laws, the law of double negation $x'' = x$ and the complementation laws $x \vee x' = 1$ and $x \wedge x' = 0$.

Note that it suffices to postulate only one of the De Morgan laws. For instance, if $(x \vee y)' = x' \wedge y'$ holds, then

$$(x \wedge y)' = (x'' \wedge y'')' = (x' \vee y')'' = x' \vee y' \,.$$

Roughly speaking, one could say that an ortholattice is a Boolean algebra without distributivity. The inclusion of Boolean algebras within ortholattices can be specified as follows.

Proposition 4.10.1. *The following conditions are equivalent for an ortholattice L:*

(i) *L is a Boolean algebra ;*

(ii) *L is distributive ;*

(iii) *L is uniquely complemented .*

PROOF: (i)$\Longleftrightarrow$(ii): By the definition of ortholattices.

(ii)$\Longrightarrow$(iii): By Proposition 1.1.

(iii)$\Longrightarrow$(i): By Theorem 8.1, condition 3. □

An *orthomodular lattice* is an ortholattice which satisfies the *orthomodular law*

OM $x \leq y \Longrightarrow x \vee (x' \wedge y) = y \,.$

Clearly OM implies that every modular ortholattice is an orthomodular lattice. But, despite the terminology, an orthomodular lattice need not be a modular lattice. Kalmbach (op.cit.), Ch.1, Exercise 20 gives an example of an orthomodular lattice which is not modular and two examples of modular ortholattices (hence orthomodular lattices) that are not distributive. One of them is the "Chinese lantern", i.e., the six-element lattice having four pairwise incomparable elements a, b, c, d with $a' = b$, $b' = a$, $c' = d$, $d' = c$, $0' = 1$, $1' = 0$.

As for the equational characterization (2.1.2) of modularity (2.1.1), it is easy that OM is equivalent to

OM1 $x \vee (x' \wedge (x \vee y)) = x \vee y$,

and also to

OM2 $(x \wedge y) \vee ((x \wedge y)' \wedge y) = y$,

therefore the orthomodular lattices form a subvariety of the equational class of ortholattices. Note also that the concepts of ortholattice and orthomodular lattice are self-dual. To see this just write OM1 or OM2 for x' and y' and take complements of both sides. Likewise, any variety of ortholattices is self-dual.

A few short independent systems of axioms for the above algebras have been given.

Thus, ortholattices were defined by Sobociński [1975b] via the system $\{O1, O2, O3, O4\}$, where

O1 $a \vee b = b \vee a$,

O2 $a \wedge (a \vee b) = a$,

O3 $a \vee (b \wedge b') = a$,

O4 $(a \vee b) \vee c = ((c' \wedge b')' \wedge a')'$,

by the simpler system $\{O2, O3, O5\}$ due to Beran [1976], where

O5 $(a \vee b) \vee c = (c' \wedge b') \vee a$,

and also by the system $\{O6, O7\}$ due to Sobociński [1979], where

O6 $(b \wedge (c \wedge a)) \vee a = a$,

O7 $((a \wedge (b \wedge (c \vee c))) \vee d) \vee e = (((((g \wedge g') \vee (c' \wedge f')') \wedge (a \wedge b)) \vee e) \vee ((h \vee d) \wedge d)$.

Orthomodular lattices are defined by systems $\{O2, O3, OM3\}$ and $\{O6, OM4\}$, given by Sobociński [1976b] and [1979], respectively, where

OM3 $a \vee ((a \vee ((b \vee c) \vee d)) \wedge a') = ((d' \wedge c')' \vee b) \vee a$,

OM4 $((((a' \wedge (a \vee ((b \vee c) \vee d))) \vee a) \wedge (m \wedge n)) \vee g) \vee e$
          $= ((n \wedge (((h' \wedge h) \vee (((d' \wedge c')' \vee b) \vee a)) \wedge m)) \vee e) \vee ((f \vee g) \wedge g)$.

Modular ortholattices are defined by systems $\{MO1, MO2, MO3\}$ and $\{O6, MO4\}$, given by Sobociński [1976a] and [1979], respectively , where

MO1 $((a \wedge b) \vee (a \wedge c)) \vee (d \wedge d') = ((c \wedge a) \vee b) \wedge a$,

MO2 $(b \vee a) \wedge a = a$,

MO3 $(a \vee b) \vee c = a \vee (b' \wedge c')'$,

MO4 $(((a \wedge (b \vee c)) \wedge (m \wedge (g \vee n))) \vee d) \vee e$

$= ((((h \wedge h') \vee (n' \wedge g')') \wedge ((a \wedge ((b \wedge (a \vee c)) \vee c)) \wedge m)) \vee e) \vee$
$((f \vee d) \wedge d)$.

We prove below two of these results. However for the independence part the reader is referred to the original papers.

Proposition 4.10.2. *((Beran [1976]) The system* $\{O2, O3, O5\}$ *characterizes ortholattices.*

PROOF: It follows from O2 with $b := b \wedge b'$ and O3, that $a \wedge a = a$. Therefore, taking $b := c := c \wedge c'$ in O5 we obtain, via O3,

(1) $$a = (c \wedge c')'' \vee a .$$

Further, taking $a := c \wedge c'$ in (1) and using again O3, we get $c \wedge c' = (c \wedge c')''$, which transforms (1) into

(2) $$a = (c \wedge c') \vee a .$$

Now O3 and (2) imply $(c \wedge c') \vee (b \wedge b') = c \wedge c'$ and $b \wedge b' = (c \wedge c') \vee (b \wedge b')$, respectively, hence $c \wedge c' = b \wedge b'$. Therefore this element is a constant, say 0, so that O3 and (2) read

(3) $$a \vee 0 = a = 0 \vee a .$$

Moreover, we have seen that $0 = 0''$. Therefore, setting $0' = 1$, we have also $0 = 1'$.

Taking $c := 0$ in O5 we obtain

(4) $$a \vee b = (1 \wedge b')' \vee a ,$$

which for $a := 0$ yields $b = (1 \wedge b')'$ and this transforms (4) into

(5) $$a \vee b = b \vee a .$$

Besides, we have also

(6) $$b \vee c = (c' \wedge b')'$$

by O5 with $a := 0$, hence $a' \vee a = (a' \wedge a'')' = 0' = 1$, so that the two complementation laws hold.

Furthermore, O2 implies $a' = a' \wedge (a' \vee a) = a' \wedge 0'$, whence using once again (6) we get $a'' = (a' \wedge 0')' = 0 \vee a = a$ and we have thus proved

(7) $$a'' = a \, .$$

Now from (7), (6) and (5) we infer

$$b \wedge c = (b'' \wedge c'')'' = (c' \vee b')' = (b' \vee c')' = (c'' \wedge b'')'' = c \wedge b \, ,$$

while from O5, (6) and (5) we infer

$$(a \vee b) \vee c = (c' \wedge b')' \vee a = (b \vee c) \vee a = a \vee (b \vee c) \, .$$

At this point it is routine to use the De Morgan law (6) together with (7) in order to prove the associativity of $\wedge$ and the duals of O2 and (3).

□

Lemmas 10.1, 10.2 and Proposition 10.3 below are due to Sobociński [1976a].

Lemma 4.10.1. MO1, MO2 *and* MO3 *imply* O3.

PROOF: It follows from MO1 and MO2 that

(8) $$((a \wedge a) \vee (a \wedge c)) \vee (d \wedge d') = ((a \wedge a) \vee a) \wedge a = a \, ,$$

then from (8) and MO2 we get

(9) $$a \wedge (d \wedge d') = (((a \wedge a) \vee (a \wedge c)) \vee (d \wedge d')) \wedge (d \wedge d') = d \wedge d' \, ,$$

whence (8), (9) and MO3 imply

$$a = ((a \wedge a) \vee (a \wedge (d \wedge d'))) \vee (d \wedge d') = ((a \wedge a) \vee (d \wedge d')) \vee (d \wedge d')$$

$$= ((a \wedge a) \vee (d \wedge d')) \vee (d \wedge d') = (a \wedge a) \vee ((d \wedge d')' \wedge (d \wedge d')')' \, ,$$

therefore setting

$$\delta = ((d \wedge d')' \wedge (d \wedge d')')' \, ,$$

we have $a = (a \wedge a) \vee \delta$. This implies $a \wedge \delta = ((a \wedge a) \vee \delta) \wedge \delta = \delta$ by MO2, hence using (8) we obtain

$$a = ((a \wedge a) \vee (a \wedge \delta)) \vee (d \wedge d') = ((a \wedge a) \vee \delta) \vee (d \wedge d') = a \vee (d \wedge d') \, .$$

□

Lemma 4.10.2. MO1, MO2 *and* MO3 *imply*

(10) $$(a \wedge b) \vee (a \wedge c) = ((c \wedge a) \vee b) \wedge a \, ,$$

(11) $$a = (a \wedge a) \vee (a \wedge c) \, ,$$

and the idempotency laws.

PROOF: Identity O3 holds by Lemma 10.1, therefore MO1 reduces to (10). From MO2 with $c := c \wedge a$ we obtain (11). Now (11) and (10) imply

$$(12) \qquad a \wedge a = ((a \wedge a) \vee (a \wedge a)) \wedge a = ((a \wedge (a \wedge a)) \vee (a \wedge a) ,$$

while from MO2 and (11) we infer

$$(13) \qquad a \wedge a = ((a \wedge a) \vee (a \wedge a)) \wedge (a \wedge a) = a \wedge (a \wedge a) ,$$

so that (11), (13) and (12) imply

$$a = (a \wedge a) \vee (a \wedge a) = (a \wedge ((a \wedge a)) \vee (a \wedge a) = a \wedge a ,$$

which implies further $a = (a \wedge a) \vee (a \wedge a) = a \vee a$, via (11). □

Proposition 4.10.3. *The system* $\{MO1, MO2, MO3\}$ *characterizes modular ortholattices.*

PROOF: First we note that it suffices to prove O2 and the commutativity laws. For commutativity reduces MO3 to O5, while O3 holds by Lemma 10.1, therefore the algebra will be an ortholattice by Proposition 10.2. Besides, Lemma 10.2 ensures that (10) holds, and this identity will coincide with the modularity condition (2.1.2') due to commutativity.

We are going to use the conclusions of Lemma 10.2. First we prove O2. It follows from MO2 and (11) that

$$(14) \qquad a \wedge c = ((a \wedge a) \vee (a \wedge c)) \wedge (a \wedge c) = a \wedge (a \wedge c) ,$$

while (10) implies

$$(15) \qquad (a \wedge b) \vee a = (a \wedge b) \vee (a \wedge a) = ((a \wedge a) \vee b) \wedge a = (a \vee b) \wedge a ,$$

and using (11), (10), (14) and (15) we get

$$(16) \qquad \begin{aligned} a = a \wedge a &= ((a \wedge a) \vee (a \wedge c)) \wedge a = (a \wedge (a \wedge c)) \vee (a \wedge a) \\ &= (a \wedge c) \vee a = (a \vee c) \wedge a , \end{aligned}$$

while using (16) twice, then (10), we obtain O2:

$$a = (a \vee (a \vee b)) \wedge a = (((a \vee b) \wedge a) \vee (a \vee b)) \wedge a$$

$$= (a \wedge (a \vee b)) \vee (a \wedge (a \vee b)) = a \wedge (a \vee b) .$$

Finally the desired commutativities are obtained as follows. From MO2 and (16), then (10) and O2 we get

$$a \vee b = ((b \vee a) \wedge a) \vee ((b \vee a) \wedge b) = ((b \wedge (b \vee a)) \vee a) \wedge (b \vee a)$$

$$= (b \vee a) \wedge (b \vee a) = b \vee a \,,$$

while (10), (15) and (16) imply

$$a \wedge b = (a \wedge b) \vee (a \wedge b) = ((b \wedge a) \vee b) \wedge a = ((b \vee a) \wedge b) \wedge a = b \vee a \,.$$

$$\square$$

In the second part of this section we prove several metatheorems in which GGP terms and a generalization of them, Mal'cev terms, play a crucial role.

Remark 4.10.1. In every orthomodular lattice

$$p(x, y, z) = (x \wedge z) \vee ((y \wedge z)' \wedge z) \vee ((x \wedge y)' \wedge x)$$

is obviously a GGP term

Theorem 4.10.1. *Every finitely based variety of orthomodular lattices is one-based.*

PROOF: By Corollary 9.4 and Remark 10.1. $\square$

Corollary 4.10.1. *The variety of orthomodular lattices is one-based.*

Furthermore, let O_8 be the ortholattice obtained by adjoining 0 and 1 to the two incomparable chains $a < b < c$ and $c' < b' < a'$. By removing c and c' one obtains an ortholattice known as O_6. Clearly O_6 is not an orthomodular lattice (OM fails for $x := a$, $y := b$) and in fact it is known that an ortholattice is orthomodular if and only if it does not include O_6 as a subalgebra; see e.g. Kalmbach [1983], Ch.1, Theorem 2. We will refer to this characterization of orthomodularity as the *forbidden-subalgebra theorem.*

Lemmas 10.3, 10.4, Theorems 10.2, 10.3 and Corollaries 10.2, 10.3 below are due to D.Kelly and Padmanabhan [2005].

Lemma 4.10.3. *The ortholattice O_8 is not congruence-permutable.*

PROOF: The principal congruences $\theta(a, b)$ and $\theta(b, c)$ satisfy $a\theta(a, b)b$ & $b\theta(a, b)c$, but $(c, a) \notin \theta(a, b) \circ \theta(b, c)$. $\square$

Theorem 4.10.2. *A variety of ortholattices is congruence-permutable if and only if it is a variety of orthomodular lattices.*

PROOF: Every two congruences of an orthomodular lattice permute; cf. Kalmbach [1983], Ch.2, Exercise 4. Conversely, let $\mathbf{V}$ be a variety of ortholattices that is congruence-permutable. Since O_8 is a subalgebra of the square of O_6, $\mathbf{V}$ does not contain O_6 by Lemma 10.3, therefore every algebra in $\mathbf{V}$ is orthomodular by the forbidden-subalgebra theorem. □

To obtain further results, we recall a theorem due to Mal'cev, which says that a variety $\mathbf{V}$ is congruence-permutable if and only if there is a term $p(x, y, z)$ such that the following identities hold in every $\mathbf{V}$-algebra:

MV1 $p(x, y, y) = x$,

MV2 $p(y, y, x) = x$.

Such a polynomial p is called a *Mal'cev term*; see e.g. Grätzer [1979], Theorem 26.4, or Burris and Sankappanavar [1981], Ch.2, Theorem 12.2. Note that every GGP term (cf. §9) is a Mal'cev term and the property of being a Mal'cev term is also relative to a variety.

Lemma 4.10.4. *Let p be a Mal'cev term for orthomodular lattices. Then orthomodular lattices are characterized by* MV1 *or* MV2 *among ortholattices according as* MV1 *or* MV2 *fails in* O_6.

PROOF: Since O_6 is not orthomodular, the variety generated by it is not congruence-permutable, by Theorem 10.2. It follows by the theorem of Mal'cev that O_6 cannot satisfy both MV1 and MV2. If MV1 fails in O_6 then any ortholattice that satisfies MV1 has not O_6 as a subalgebra, hence it is orthomodular by the forbidden-subalgebra theorem. The other case is treated similarly. □

Theorem 4.10.3. *An ortholattice is orthomodular if and only if it has a* GGP-*term.*

PROOF: The "only if" part is Remark 10.1. Conversely, an ortholattice having a GGP term is orthomodular by Lemma 10.4. □

Corollary 4.10.2. *If p is a ternary term symmetric in x an z and such that* MV1 *holds in orthomodular lattices, then condition* MV1 *characterizes orthomodular lattices among ortholattices.*

PROOF: Conditions MV1 and MV2 are now equivalent, therefore p is a Mal'cev term and the desired conclusion follows by Lemma 10.4. □

Corollary 10.2 is a generator of several relative characterizations of orthomodular lattices. Here is an example:

Corollary 4.10.3. *Identity*

$$(x \wedge y) \vee ((x \wedge y)' \wedge x) = x$$

characterizes orthomodular lattices among ortholattices.

PROOF: By Corollary 10.2 and Remark 10.1.   □

The paper by D. Kelly and Padmanabhan [2003] provides several results in this line. The main one is that for every $n \geq 2$, every finitely based variety of orthomodular lattices has an independent self-dual n-basis.

The following companion of Theorem 10.3 is also due to D.Kelly and Padmanabhan [2007]:

Theorem 4.10.4. *An orthomodular lattice is a Boolean algebra if and only if it has a unique GGP term.*

PROOF: The "if" part is Proposition 9.4. To prove the "only if" part, note that Remark 10.1 and its dual provide two GGP terms. According to the hypothesis, they must coincide, so that identity $B_2^{\vee\wedge}$ holds, which characterizes Boolean algebras by Proposition 2.4.   □

Several GGP terms for orthomodular lattices have been pointed out in the literature. In a Boolean algebra they coincide with the unique term described in Proposition 9.4. Alternatively, this can be easily checked by direct computation in Boolean algebra.

Chapter 5

Further Topics and Open Problems

In this short chapter we present, in a rather informal and sketchy way, a few lines of research related to the material of this book and leading to several open problems.

I. Tarski-type theorems on independent equational bases

The topic of this monograph belongs to what is widely known as equational logic and equational theories of algebras. We may safely say that Alfred Tarski School at the University of California, Berkeley (during the late 1960's) was, in a sense, the Mecca for equational logic in modern times. Let **K** be a finitely-based equational theory of algebras. Following Tarski [1968], let $\nabla(\mathbf{K})$ denote the equational spectrum of **K**, that is, the set of cardinalities of independent equational bases of **K**. Tarski has shown that if **K** satisfies an equation of the form $f = x$, where f has at least two occurrences of the variable x, then $\nabla(\mathbf{K})$ is an unbounded interval. This is what we call Tarski's Unbounded Theorem (TUT). McNulty proved a stronger version of TUT in 1976. For such varieties, Tarski has further shown that if i, j belong to the set $\nabla(\mathbf{K})$, then every integer between i and j also belongs to $\nabla(\mathbf{K})$. This we call Tarski's Interpolation Theorem (TIT).

For a variety admitting a duality , let $\nabla_{sd}(\mathbf{K})$ be the set of all natural numbers n such that **K** has an independent *self-dual* basis with n identities. First we give a rather simple example of a finitely-based self-dual variety which demonstrates the failure of the self-dual analogue of TIT. Indeed, if **K** is the variety of all algebras of type (2,2) with two semilattice operations, then it is easy to show that 2, 4 belong to $\nabla_{sd}(\mathbf{K})$. However, it has been shown (D. Kelly and Padmanabhan [2004]) that **K** *cannot* be defined by any independent self-dual set of three identities. Thus, the self-dual analogue of Tarski's Interpolation Theorem is not true if we insist that the equational basis in question enjoys some additional syntactic property (e.g., being self-dual).

However, the self-dual analogues of TIT and TUT are valid for all "nice" lattice varieties **K** (including distributive lattices, modular lattices, Boolean algebras, ortholattices, OML's and any finitely-based self-dual variety of lattices). In fact, all these self-dual equational theories have independent self-dual bases for all $n \geq 2, 3$ or 4. For example, in D. Kelly and Padmanabhan [2004] it has been shown that

$$\nabla_{sd}(\mathbf{K}) = [1, \omega) \text{ if } \mathbf{K} \text{ is the self-dual variety of all one-element latices,}$$
$$= [2, \omega) \text{ if } \mathbf{K} \text{ is the self-dual variety of all lattices,}$$
$$= [3, \omega) \text{ if } \mathbf{K} \text{ satisfies a self-dual identity } f = df,$$
$$= [4, \omega) \text{ otherwise.}$$

Tarski's original 1975 proof (without the further syntactic restriction that the equational bases be self-dual) uses the concept of closure system and is existential in nature. In sharp contrast, D.Kelly and Padmanabhan exhibit the actual self-dual bases by means of a natural blow-up process based on elementary number-theoretic models. Thus, these independent equational bases may be construed as the first constructive proof of Tarski's original theorems as well.

Open problems

1. Let **K** be the variety of all algebras of type (2,2) with two semilattice operations. As mentioned above, 2 and 4 belong to $\nabla_{sd}(\mathbf{K})$, but not 3. Does 5 belong to $\nabla_{sd}(\mathbf{K})$? In other words, is it possible to define the semilattice properties of two operations by an independent self-dual set of five identities ?

2. Characterize the set $\nabla_{sd}(\mathbf{K})$.

3. Is there an example of a variety **K** of algebras admitting a duality but having no independent self-dual bases ?

II. Huntington varieties of lattices

Unlike the classical algebraic theories like groups and rings, substructures play a rather unique role in the equtional theory of lattices. Just recall the famous 1897 theorem of Dedekind that a lattice is modular if and only if it contains no pentagon (Theorem 2.1.1). As early as 1943, Lowig gave the first example of a "forbidden" non-self-dual lattice, which was published in the *Annals of Mathematics* (we don't see any research paper in lattice theory published nowadays in such main-stream journals, do we?). That subdirectly irreducible lattice has nine elements and is, in fact, one of the covers of N_5. But its publication was way ahead of its time and did not catch the atention of many lattice theorists. The lattice-theory world had to wait for the landmark discoveries of bounded homomorphisms and splitting lattices by McKenzie (see Ch.2 in Freese, Jezek and Nation [1995]) and the Magic Lemma of Jónsson along with the Jónsson terms to fully appreciate the interplay between equational theory of lattices and internal algebraic structure of lattices. Recall that L is a *splitting lattice* if there is a lattice identity $f = g$ (the *conjugate identity* of the lattice L) such that a lattice K satisfies $f = g$ if and only if L is not in the variety generated by the lattice K. McKenzie characterized splitting lattices as those finite subdirectly irreducible lattices which are bounded epimorphic images of a free lattice. These lattices are *semidistributive* (i.e., they satisfy $SD^\vee$ and $SD^\wedge$; see Theorem 4.8.3). Moreover, every finite splitting lattice satisfies an identity which formally implies these semidistributive implications. It was this aspect of lattice theory that we exploited in Section 8 of Chapter 4 to discover infinitely many Huntington varieties of lattices, i.e., lattice varieties in which every uniquely complemented lattice is distributive.

Open problems

4. We conjecture there exists no largest Huntington variety.

5. We conjecture that if **A** and **B** are Huntington varieties, then so is **A** ∨ **B**. In support of this conjecture we have shown that **M** ∨ N_5 is Huntington; cf. Corollary 8.8.1.

Note that the analogue of Problem 5 for infinite joins has a negative answer. For, using the celebrated Jónsson Lemma, one can show that every splitting lattice generates a non-modular Huntington variety. It follows that the infinite join of all these Huntington varieties is the variety **L** of all lattices, because in fact, according to a well-known result of Alan Day, the class of all splitting lattices generates **L**. But trivially the variety **L** is not Huntington. As a matter of fact, **L** is a Dilworth variety; cf. Problem 9.

6. Call a lattice variety **K** *strictly nonmodular* if **K** does not contain M_3. We conjecture that every strictly nonmodular variety is Huntington.

7. Prove or disprove that the variety generated by a finite lattice is Huntington.

8. Prove or disprove that if a finite lattice L generates a Huntington variety, then so does $L[d]$ obtained by Day's doubling construction. Here $L[d]$ is obtained by "doubling the element d"; see Day [1977] and Jipsen and Rose [1992], page 88, Lemma 4.11.

Lemma 4.11 proves that if L is a finite distributive lattice, then $L[d]$ is almost distributive – in particular, it is semidistributive and hence by the results mentioned in Chapter 4, the variety generated by $L[d]$ is Huntington. Our problem asks whether this is always the case, if L is not necessarily distributive.

9. A lattice variety **K** is called a *Dilworth variety* if every lattice in **K** can be embedded in some uniquely complemented lattice in **K**. For instance, the variety of all lattices is Dilworth. Adam and Sichler proved that every p–modular lattice can be embedded in a uniquely complemented p–modular lattice. Is there an example of a lattice variety that is neither Huntington nor Dilworth ?

10. Project. Carry out a similar program for the bi-complemented lattices of Chen and Grätzer [1969]. For example, characterize semidistributive complemented lattices in which every element has precisely two complements.

11. Day, Nation and Tschantz [1989] have introduced the implication of semidistributivity: $x \wedge y = x \wedge z$ & $z \vee x = z \vee y \implies x \leq y$. We conjecture that if a lattice variety satisfies this implication, then it is a Huntington variety.

III. Binary reducts of Boolean algebras

In Chapter 4 we discussed several avatars of Boolean algebras and proved that each one of them can be defined by a single axiom. This was an easy consequence of the fact that the variety of Boolean algebras is arithmetical. However, most of the reducts of Boolean algebras do not have the arithmetical property. In theory there are sixteen binary reducts of Boolean algebras, each one corresponding to a binary term – an element of the free Boolean algebra on two generators. By symmetry and duality there are just eight essentially distinct binary reducts: (B, f), where

$$f(x, y) \in \{0, x, x', x \wedge y, x \wedge y', x' \vee y, x' \wedge y', (x \wedge y') \vee (x' \wedge y)\} .$$

Open problems

12. Project. Find minimal equational bases for the varieties generated by (B, f).

The following minimal equational bases are already known:

for $f(x, y) = x$ (left semigroups): $f(x, y) = x$;

for $f(x, y) = x \wedge y$ (semilattices): two identities; cf. Chapter 1, §1 (these systems are minimal by Theorem 1.3.1);

for $f(x, y) = x' \wedge y'$ (Sheffer stroke): single identity; cf. Chapter 4, $\mathrm{B}^!_{87}$;

for $f(x, y) = x' \vee y$ (implication algebras): two identities; see Theorems 5.3.1 and 5.3.2 below.

13. We conjecture there is no single identity defining Boolean algebras shorter than $\mathrm{B}^!_{87}$ among all possible single identities (i.e., not only among those expressed in terms of the Sheffer stroke).

The implicational fragment of the two-element Boolean algebra is the class of algebras of type (2) having a single binary operation $\to$ with the interpretation that $x \to y = x' \vee y$. Abbott [1967] first defined these *implication algebras* by the following three identities:

(1)
$$(x \to y) \to x = x \, ,$$

(2)
$$(x \to y) \to y = (y \to x) \to x \, ,$$

(3)
$$x \to (y \to z) = y \to (x \to z) \, .$$

It is natural to look for a minimal equational basis of implication algebras. This problem is solved by Theorems 5.3.1 and 5.3.2 below, due to Gareau and Padmanabhan (unpublished). We are going to use the well-known fact that the relation $x \leq y \iff x \to y = 1$ makes an implication algebra into a poset with greatest element 1, where $1 = z \to z$ for any z. This implies the identity $x \to 1 = 1$. Note also that implication algebras with 0 coincide with Boolean algebras.

Lemma 5.3.1 *Let $f(x_1, \ldots, x_n)$ be a non-constant term function of an implication algebra L. If L satisfies an identity of the form $f(x_1, \ldots, x_n) = x_i$, where $1 \leq i \leq n$, then x_i is the last variable which occurs in f.*

Comment: In particular L may be a Boolean algebra.

PROOF: This follows from the identity $x_n \leq f(x_1, \ldots, x_n)$, where x_n stands for the last variable which appears in f. We prove the identity by induction on the length of the expression of f. For $f := x \to y$ we note that (3) implies

$$y \to (x \to y) = x \to (y \to y) = x \to 1 = 1 \, ,$$

i.e., $y \leq x \to y$. The inductive step follows from $x_n \leq f_2 \leq f_1 \to f_2 = f$.

□

Theorem 5.3.1 *Implication algebras are not one-based.*

PROOF: Suppose $f(x_1, \ldots, x_n) = g(x_1, \ldots, x_n)$ is a single axiom for implication algebras. If none of the term functions f, g reduces to a single variable, then the constant algebra $x \to y = 1$ would be a model for that axiom; but this algebra does not satisfy axiom (1). For the same reason the axiom cannot be of the form $f(x_1, \ldots, x_n) = 1$ either. Therefore the single axiom is of the form $f(x_1, \ldots, x_n) = x_i$, where $1 \leq i \leq n$.

Now it follows by Lemma 5.3.1 that the single identity is $f(x_1, \ldots, x_n) = x_n$. Then the right projection $x \to y = y$ is a model for the single axiom, but it does not fulfil axiom (2).   □

Theorem 5.3.1 can be extended to orthomodular lattices in the following way. As shown by Kalmbach [1983], Ch.4, §15, Theorem 3, in every orthomodular lattice there are exactly five term functions $\to_j$ with the property

$$x \leq y \Longleftrightarrow x \to_j y = 1 \, .$$

If the OML is a Boolean algebra, all of them reduce to $x \to y = x' \vee y$.

Corollary 5.3.1 *The equational theory of* $(L, \to_j)$ $(1 \leq j \leq n)$ *has no single axiom.*

PROOF: Suppose one of the five implications $\to_j$ is characterized by a single axiom. As in the proof of Theorem 5.3.1, the axiom must be of the form $f(x_1, \ldots, x_n) = x_i$. But this axiom is fulfilled by the implication $x \to y = x' \vee y$ of a Boolean algebra. So it follows by Lemma 5.3.1 that $i = n$ and again the right projection is a model of the axiom but fails to satisfy the commutativity of the disjunction operation of the OML, which is expressed by an identity $\varphi(x, y) = \psi(x, y)$, where the last variables in φ and ψ are x and y.   □

Theorem 5.3.2 *The identities* (1) *and*

$$(4) \qquad\qquad x \to (y \to ((z \to u) \to u)) = y \to (x \to ((u \to z) \to z))$$

form a basis for the equational theory of implication algebras.

PROOF: Note first that axioms (3) and (2) imply (4). The following proof of the converse implication is based on an automated proof provided by Prover9.

Setting $u := z$ in (4) and taking into account (1) we obtain (3). So it remains to prove (2).

By applying (1) twice we obtain

$$(5) \qquad x \to (x \to y) = ((x \to y) \to x) \to (x \to y) = x \to y \ .$$

By applying (4) with

$$x := (((y \to x) \to x) \to z), \ y := u, \ z := x, \ u := y \ ,$$

then (1) with $x := (y \to x) \to x$, $y := z$, we obtain

$$(6) \qquad \begin{aligned} (((y \to x) \to x) \to z) &\to (u \to ((x \to y) \to y)) \\ = u &\to ((((y \to x) \to x) \to z) \to ((y \to x) \to x) \\ = u &\to ((y \to x) \to x) \ . \end{aligned}$$

Further, by using in turn (3), (1), (6) with $u := ((x \to y) \to y) \to u$, and (3), we obtain

$$(x \to y) \to ((((y \to x) \to x) \to z) \to y)$$

$$= (((y \to x) \to x) \to z) \to ((x \to y) \to y)$$

$$= (((y \to x) \to z) \to ((((x \to y) \to u) \to ((x \to y) \to y))$$

$$= (((x \to y) \to y) \to u) \to ((y \to x) \to x)$$

$$= (y \to x) \to ((((x \to y) \to y) \to u) \to x) \ ,$$

hence

$$(7) \qquad \begin{aligned} (x \to y) &\to ((((y \to x) \to x) \to z) \to y) \\ = (y \to x) &\to ((((x \to y) \to y) \to u) \to x) \ . \end{aligned}$$

On the other hand, using the abbreviation $((y \to x) \to x) \to z = u$ and applying in turn (3), (5), (6) and (1), we obtain

$$(x \to y) \to (u \to y) = u \to ((x \to y) \to y) = u \to (u \to ((x \to y) \to y))$$

$$= (((y \to x) \to x) \to z) \to ((y \to x) \to x) = (y \to x) \to x \ ,$$

that is,

$$(8) \qquad (x \to y) \to ((((y \to x) \to x) \to z) \to y) = (y \to x) \to x \ .$$

By interchanging x and y we get

(8′) $(y \to x) \to ((((x \to y) \to y) \to z) \to x) = (x \to y) \to y$;

but the left-hand sides of (8) and (8′) coincide by (7), therefore $(y \to x) \to x = (x \to y) \to y$, which is (2).   □

Historic remark The fact that this equational theory is 2-based was first proved by Meredith and Prior [1968], who gave a system consisting of equations (1) and

(9) $(x \to y) \to (z \to y) = (y \to x) \to (z \to x)$.

Corollary 5.3.2 *The complete equational spectrum for this variety is* $[2, \omega)$, *i.e., for all* $n \geq 2$ *there exists an independent n-basis for implication algebras.*

PROOF: Theorem 8 in the paper by Tarski [1968] says that if Θ is an equational theory, τ is a term, x is a variable with at least two occurrences in τ such that the equation $\tau = x$ belongs to Θ and $0 < n \in \nabla(\Theta)$, then $[n, \omega) \subseteq \nabla(\Theta)$. In view of the above theorems, we can apply Tarski's theorem with $\tau := (x \to y) \to x$ and $n := 2$.   □

The above corollary can be further extended. Consider, for instance, the Boolean *subtraction* function $x - y = x \wedge y'$, which is the dual of $y \to x$ and can be referred to as the *mirror image* or *skew dual* of the implication. Kalman [1960] characterized the equational theory of algebras $(L, -)$ satisfying all the identities true for the Boolean function $x \wedge y'$ by the system

(10) $x - (y - x) = x$,

(11) $x - (x - y) = y - (y - x)$,

(12) $(x - y) - z = (x - z) - (y - z)$.

(Incidentally, (10) and (11) are the mirror images of (1) and (2), respectively.) Kalman called these algebras by the name of *flocks*.

Corollary 5.3.3 *The complete equational spectrum for the Kalman variety of flocks is* $[2, \omega)$.

PROOF: From Corollary 5.3.2 by skew duality.   □

IV. Frink-type theorems for varieties of complemented lattices

Quoting G. Birkhoff, O. Frink gave a remarkably brief and elegant proof of the well-known theorem that every Boolean algebra is isomorphic with an algebra of sets. The novelty of the approach was that it was based on a simple bi-implication capturing the meaning of the order relation in Boolean algebras: $x \wedge y = x \Longleftrightarrow x \wedge y' = 0$. The profound direction is, of course, the sufficiency of $x \wedge y' = 0$ for $x \wedge y = x$ and this characterizes Boolean algebras among all semilattices with an extra unary operation.

More generally, let **K** be a variety of complemented lattices. A Frink-type theorem for **K** is a statement of the form: a semilattice with a unary operation belongs to **K** iff it satisfies the implication

$$f(x, y) = g(x, y) \Longrightarrow x \wedge y = x \,,$$

where f and g are terms of type (2,1). Apart from generalizing the original Frink theorem on Boolean algebras, every such result will give a new "equational expression" for the order relation $x \wedge y = x$ which is rather unique to the particular equational theory in question. For example, Adam Gareau, a graduate student at the University of Manitoba, has proved such an analogue for the class of orthomodular lattices:

$(L, \wedge, ', 0)$ is an OML if and only if $(x' \wedge (x' \wedge y)')' = y \Longrightarrow x \wedge y = x \,.$

Open problems

14. Project. Discover new Frink-type theorems for well-known varieties of complemented lattices, e.g. ortholattices, modular ortholattices, Stone lattices and Newmann algebras.

In the case of orthomodular lattices, such a theorem would have obvious implications (no pun intended!) to the various implications that can be defined in the context of the logic of quantum mechanics.

In Problems 15–22 below $(L, \vee, \wedge, ')$ is an orthomodular lattice and $f(x, y)$ is a binary term of L. Problems 16–20 are conjectures.

15. Find all binary reducts (L, f) of L which have a single axiom for their equational theory.

This problem is decidable, because the free OML on two generators is finite; cf. Kalmbach [1983], p. 218.

16. If the equational theory of the binary reduct (L, f) is not one-based, then it is two-based.

17. If (L, f) is cancellative, then (L, f) is an Abelian group.

18. If (L, f) is a cancellative semigroup, then (L, f) is an Abelian group.

19. If (L, f) is a group, then $(L, \vee, \wedge,')$ is a Boolean algebra.

20. If (L, f) is cancellative, then the lattice reduct $(L, \vee, \wedge)$ is distributive.

21. Characterize the variety generated by $(L, x \wedge y')$ in OML's (i.e., analogue of Kalman's theorem for orthomodular lattices).

22. Project. Same problems as above for the five implications defined in an OML ; see Kalmbach [1983].

Appendix A: Some Prover9 Proofs

by W. McCune

This Appendix contains several examples of lattice theory proofs found by the program Prover9 (version March-2007); cf. McCune [1]. For each example, the Prover9 input file is given, followed by the proof produced by Prover9.

Prover9 proves theorems by contradiction, and the conclusion of the theorem is assumed to be false. For example, if the conclusion is that the meet operation is commutative and associative, the input file contains the following assumption (!= is negated equality, and | is disjunction).

```
A ^ B != B ^ A | (A ^ B) ^ C != A ^ (B ^ C).
```

The terms A, B, C are constants, and terms x, y, z, u, v, w are variables. Each step of the proof contains a justification on the right-hand side listing the lines from which the step was derived.

Sholander's 2-Basis for Distributive Lattices

Here is a proof of Sholander's Theorem 3.2.1.

Input file:

```
formulas(assumptions).

% Sholander's 2-basis for distributive lattices:

x ^ (x v y) = x              # label(Absorb_A).
x ^ (y v z) = (z ^ x) v (y ^ x)   # label(Sholander).

% Denial of a standard 6-basis for Lattice Theory.

A ^ B != B ^ A |
```

```
(A ^ B) ^ C != A ^ (B ^ C) |
A ^ (A v B) != A |
A v B != B v A |
(A v B) v C != A v (B v C) |
A v (A ^ B) != A.
```

end_of_list.

Prover9 proof in 1 second:

1	$x \wedge (x \vee y) = x$ # Absorb_A	[assumption]
2	$x \wedge (y \vee z) = (z \wedge x) \vee (y \wedge x)$ # Sholander	[assumption]
3	$(x \wedge y) \vee (z \wedge y) = y \wedge (z \vee x)$	[2]
4	$A \wedge B \neq B \wedge A \mid (A \wedge B) \wedge C \neq A \wedge (B \wedge C) \mid A \wedge (A \vee B) \neq A \mid$	
	$A \vee B \neq B \vee A \mid (A \vee B) \vee C \neq A \vee (B \vee C) \mid A \vee (A \wedge B) \neq A$	
		[assumption]
5	$B \wedge A \neq A \wedge B \mid (A \wedge B) \wedge C \neq A \wedge (B \wedge C) \mid B \vee A \neq A \vee B \mid$	
	$(A \vee B) \vee C \neq A \vee (B \vee C) \quad A \vee (A \wedge B) \neq A$	[4,1]
6	$(x \wedge y) \wedge (y \wedge (z \vee x)) = x \wedge y$	[3,1]
7	$(x \wedge y) \vee (y \wedge y) = y$	[3,1]
8	$x \vee (y \wedge (x \vee z)) = (x \vee z) \wedge (y \vee x)$	[1,3]
9	$(x \wedge (y \vee z)) \vee y = (y \vee z) \wedge (y \vee x)$	[1,3]
10	$((x \wedge y) \vee (z \wedge y)) \vee (u \wedge (z \vee x)) = (z \vee x) \wedge (u \vee y)$	[3,3]
11	$x \wedge ((x \wedge y) \wedge (z \vee x)) = x$	[1,6,1]
12	$(x \wedge y) \wedge y = x \wedge y$	[1,6]
13	$(x \wedge y) \wedge ((y \wedge (z \vee x)) \wedge (u \vee (x \wedge y))) = x \wedge y$	[6,6,6]
14	$x \vee ((x \vee y) \wedge (x \vee y)) = x \vee y$	[1,7]
15	$((x \wedge y) \vee (z \wedge y)) \vee ((z \vee x) \wedge (z \vee x)) = z \vee x$	[3,7]
16	$(x \vee y) \wedge (x \vee x) = x \vee x$	[1,8]
17	$(x \wedge (y \vee y)) \vee (y \vee y) = y \vee y$	[8,7,1]
18	$x \vee (y \wedge (x \vee z)) = (x \vee z) \wedge ((x \vee z) \wedge (x \vee y))$	[12,8,9]
19	$x \vee (x \vee x) = x \vee x$	[8,14,1]
20	$(x \wedge x) \vee x = x$	[7,19,7]
21	$(x \wedge y) \vee ((x \wedge x) \wedge y) = y \wedge x$	[20,3]
22	$x \wedge ((x \vee y) \wedge x) = x$	[20,11]
23	$x \wedge (((x \vee y) \wedge x) \wedge (z \vee x)) = x$	[22,6,22]
24	$x \vee (((x \vee y) \wedge x) \wedge ((x \vee y) \wedge x)) = (x \vee y) \wedge x$	[22,7]
25	$(x \vee x) \wedge (y \wedge (x \vee x)) = (x \vee x) \wedge (y \vee (x \vee z))$	[16,3]
26	$(x \vee x) \wedge ((x \vee y) \vee z) = x \vee x$	[16,3,17]
27	$(x \vee x) \wedge ((x \vee x) \wedge (y \vee (x \vee z))) = x \vee x$	[16,6,16]
28	$x \wedge ((y \wedge x) \vee (y \wedge x)) = (y \wedge x) \vee (y \wedge x)$	[7,16]
29	$((x \wedge x) \vee y) \wedge x = x$	[7,16,7]
30	$x \wedge x = x$	[20,16,7,7]
31	$(x \vee y) \wedge x = x$	[29,30]
32	$x \vee x = x$	[24,31,31,30,31]

33	$x \wedge y = y \wedge x$	[21,30,32]
34	$((x \wedge y) \vee (z \wedge y)) \vee (z \vee x) = z \vee x$	[15,30]
35	$(x \wedge y) \vee y = y$	[7,30]
36	$x \wedge (x \wedge (y \vee x)) = x$	[23,33,1]
37	$x \wedge (y \wedge x) = y \wedge x$	[28,32,32]
38	$x \wedge (x \wedge (y \vee (x \vee z))) = x$	[27,32,32,32]
39	$x \wedge ((x \vee y) \vee z) = x$	[26,32,32]
40	$x \vee (y \wedge x) = x \wedge (y \vee (x \vee z))$	[25,32,32,32]
41	$C \wedge (A \wedge B) \neq A \wedge (B \wedge C) \quad \mid \quad B \vee A \neq A \vee B \quad \mid \quad (A \vee B) \vee C$	
	$\neq A \vee (B \vee C) \quad \mid \quad A \vee (A \wedge B) \neq A$	[5,33,33]
42	$(x \wedge y) \vee (z \wedge x) = x \wedge (z \vee y)$	[33,3]
43	$(x \wedge y) \wedge (x \wedge (z \vee y)) = y \wedge x$	[33,6]
44	$(x \wedge y) \vee x = x$	[33,35]
45	$(x \vee (y \wedge x)) \vee y = y \vee x$	[10,30,30,1]
46	$x \wedge (x \wedge y) = x \wedge y$	[44,1,33]
47	$(x \wedge y) \wedge (x \wedge (z \vee (x \wedge y))) = x \wedge y$	[44,11]
48	$x \wedge (y \vee (x \vee z)) = x$	[38,46]
49	$x \wedge (y \vee x) = x$	[36,46]
50	$x \vee (y \wedge (x \vee z)) = (x \vee z) \wedge (x \vee y)$	[18,46]
51	$x \vee (y \wedge x) = x$	[40,48]
52	$x \vee y = y \vee x$	[45,51]
53	$x \vee (x \wedge y) = x$	[44,52]
54	$C \wedge (A \wedge B) \neq A \wedge (B \wedge C) \quad \mid \quad C \vee (A \vee B) \neq A \vee (B \vee C)$	[41,52,52,53]
55	$x \vee (y \wedge (z \vee x)) = (z \vee x) \wedge (x \vee y)$	[49,3,52]
56	$(x \wedge y) \vee (z \wedge y) = y \wedge (x \vee z)$	[52,3]
57	$(x \wedge (y \vee z)) \vee (z \vee y) = z \vee y$	[34,56]
58	$(x \wedge y) \wedge (z \vee y) = x \wedge y$	[37,6,46,37]
59	$x \wedge ((y \vee x) \vee z) = x$	[52,39]
60	$(x \wedge y) \wedge (z \vee x) = x \wedge y$	[46,6,46,46]
61	$x \wedge (y \vee (z \vee x)) = x$	[52,48]
62	$(x \wedge y) \wedge (x \vee z) = x \wedge y$	[53,59]
63	$(x \wedge y) \wedge (x \wedge (y \wedge (z \vee x))) = x \wedge y$	[53,13,33]
64	$(x \vee y) \vee (z \wedge y) = x \vee y$	[58,51]
65	$x \wedge (y \wedge (z \wedge x)) = y \wedge (z \wedge x)$	[51,58,33]
66	$x \wedge (y \wedge (x \wedge z)) = y \wedge (x \wedge z)$	[53,58,33]
67	$(x \vee y) \vee (y \wedge z) = x \vee y$	[60,51]
68	$x \vee (y \wedge (x \wedge z)) = x$	[53,64,53]
69	$x \wedge (y \wedge (z \wedge (x \wedge u))) = y \wedge (z \wedge (x \wedge u))$	[68,58,33]
70	$((x \wedge y) \wedge z) \wedge (x \wedge (u \vee y)) = (x \wedge y) \wedge z$	[42,62]
71	$(x \wedge (y \wedge z)) \wedge (x \wedge (u \vee y)) = (y \wedge z) \wedge x$	[67,43]
72	$(x \wedge y) \wedge (x \wedge (y \wedge z)) = x \wedge (y \wedge z)$	[68,47,33]
73	$x \wedge (y \wedge (z \vee x)) = x \wedge y$	[63,72]
74	$x \wedge ((y \vee x) \wedge z) = x \wedge z$	[33,73]
75	$(x \wedge y) \wedge (y \wedge z) = (x \wedge y) \wedge z$	[51,74]
76	$(x \wedge y) \wedge (x \wedge z) = (x \wedge y) \wedge z$	[53,74]

77	$(x \wedge y) \wedge z = x \wedge (y \wedge z)$	[72,76,75]
78	$x \wedge (y \wedge (z \wedge (u \vee x))) = x \wedge (y \wedge z)$	[71,77,77,69,77]
79	$x \wedge (y \wedge z) = z \wedge (x \wedge y)$	[70,77,77,77,78,65,77]
80	$x \wedge (y \wedge z) = y \wedge (x \wedge z)$	[76,77,66,77]
81	$C \vee (A \vee B) \neq A \vee (B \vee C)$	[54,79,33,80]
82	$(x \vee y) \wedge (x \vee (y \vee z)) = x \vee y$	[48,50,33]
83	$(x \vee y) \wedge (x \vee (z \vee y)) = x \vee y$	[61,50,33]
84	$(x \vee y) \wedge (y \vee (x \vee z)) = y \vee x$	[52,82]
85	$(x \vee y) \vee (z \vee x) = (x \vee y) \vee z$	[82,55,84]
86	$(x \vee y) \vee (z \vee y) = (x \vee y) \vee z$	[83,55,84]
87	$(x \vee y) \vee z = (z \vee y) \vee x$	[83,57,85,86]
88	$(x \vee y) \vee z = x \vee (z \vee y)$	[87,52]
89	$\square$	[88,81,52,52]

Self-dual 2-Basis for Lattice Theory

Here we prove Theorem 1.2.5.

Input file:

```
formulas(assumptions).

% Self-dual (independent) 2-basis for Lattice Theory

(((x v y) ^ y) v (z ^ y)) ^ (u v ((v v y) ^ (y v w))) = y\\
                                        # label(A).
(((x ^ y) v y) ^ (z v y)) v (u ^ ((v ^ y) v (y ^ w))) = y\\
                                        # label(Dual_A).

% Denial of McKenzie's 4-baiss for Lattice Theory.

A v (B ^ (A ^ C)) != A |
     A ^ (B v (A v C)) != A |
     ((B ^ A) v (A ^ C)) v A != A |
     ((B v A) ^ (A v C)) ^ A != A.

end_of_list.
```

Prover9 proof in 117 seconds:

1. $(((x \vee y) \wedge y) \vee (z \wedge y)) \wedge (u \vee ((v \vee y) \wedge (y \vee w))) = y$ # A   [assumption]
2. $(((x \wedge y) \vee y) \wedge (z \vee y)) \vee (u \wedge ((v \wedge y) \vee (y \wedge w))) = y$
Dual_A   [assumption]

3    $A \vee (B \wedge (A \wedge C)) \neq A \mid A \wedge (B \vee (A \vee C)) \neq A \mid$
     $((B \wedge A) \vee (A \wedge C)) \vee A \neq A \mid ((B \vee A) \wedge (A \vee C)) \wedge A \neq A$
                                                [assumption]

4    $(((x \vee (y \vee ((z \vee u) \wedge (u \vee v)))) \wedge (y \vee ((z \vee u) \wedge (u \vee v)))) \vee u) \wedge (w \vee$
     $((v6 \vee (y \vee ((z \vee u) \wedge (u \vee v)))) \wedge ((y \vee ((z \vee u) \wedge (u \vee v))) \vee v7)))$
     $= y \vee ((z \vee u) \wedge (u \vee v))$ [1,1]

5    $(((x \vee (y \wedge z)) \wedge (y \wedge z)) \vee (u \wedge (y \wedge z))) \wedge (v \vee z) = y \wedge z$ [1,1]

6    $(((x \vee (y \wedge ((z \wedge u) \vee (u \wedge v)))) \wedge (y \wedge ((z$
     $\wedge u) \vee (u \wedge v)))) \vee (w \wedge (y \wedge ((z \wedge u) \vee (u \wedge v))))) \wedge$
     $(v6 \vee (u \wedge ((y \wedge ((z \wedge u) \vee (u \wedge v))) \vee v7))) = y \wedge ((z \wedge u) \vee (u \wedge v))$ [2,1]

7    $(((x \vee (y \wedge z)) \wedge (y \wedge z)) \vee (u \wedge (y \wedge z))) \wedge z = y \wedge z$ [2,1]

8    $(((x \wedge (y \vee z)) \vee (y \vee z)) \wedge (u \vee (y \vee z))) \vee z = y \vee z$ [1,2]

9    $(((x \wedge (y \vee z)) \vee (y \vee z)) \wedge (u \vee (y \vee z))) \vee (v \wedge z) = y \vee z$ [2,2]

10   $(((x \vee (y \vee z)) \wedge (y \vee z)) \vee (u \wedge z)) \wedge (v \vee ((w \vee (y \vee z)) \wedge ((y \vee z) \vee v6)))$
     $= y \vee z$ [1,4,1,1,1,1]

11   $(((x \wedge y) \vee (z \vee y)) \wedge (u \vee (z \vee y))) \vee (v \wedge ((w \wedge (z \vee y)) \vee ((z \vee y) \wedge v6)))$
     $= z \vee y$ [5,2]

12   $(((x \wedge y) \vee y) \wedge (z \vee y)) \vee (u \wedge (y \wedge v)) = y$ [5,2]

13   $(((x \vee (y \wedge (z \wedge ((u \wedge v) \vee (v \wedge w))))) \wedge (y \wedge (z \wedge ((u \wedge v) \vee (v \wedge w))))) \vee$
     $(v6 \wedge (y \wedge (z \wedge ((u \wedge v) \vee (v \wedge w)))))) \wedge v = y \wedge (z \wedge ((u \wedge v) \vee (v \wedge w)))$
                                                   [2,5]

14   $(((x \wedge y) \vee (z \vee y)) \wedge (u \vee (z \vee y))) \vee y = z \vee y$ [5,8]

15   $(((x \wedge (((y \vee z) \wedge z) \vee (u \wedge z))) \vee (((y \vee z) \wedge z) \vee (u \wedge z))) \wedge (v \vee$
     $(((y \vee z) \wedge z) \vee (u \wedge z)))) \vee (w \wedge z) = ((y \vee z) \wedge z) \vee (u \wedge z)$ [1,12]

16   $(((x \wedge (y \wedge ((z \wedge u) \vee (u \wedge v)))) \vee (y \wedge ((z \wedge u) \vee (u \wedge v)))) \wedge u)$
     $\vee (w \wedge ((y \wedge ((z \wedge u) \vee (u \wedge v))) \wedge v6)) = y \wedge ((z \wedge u) \vee (u \wedge v))$ [2,12]

17   $((x \wedge (y \wedge (x \wedge z))) \vee (u \wedge (y \wedge (x \wedge z)))) \wedge (x \wedge z) = y \wedge (x \wedge z)$ [12,7]

18   $((x \wedge (y \wedge (x \wedge z))) \vee (u \wedge (y \wedge (x \wedge z)))) \wedge (v \vee (x \wedge z)) = y \wedge (x \wedge z)$ [12,5]

19   $(((x \vee (y \wedge (z \wedge (u \wedge v)))) \wedge (y \wedge (z \wedge (u \wedge v)))) \vee (w \wedge (y \wedge (z \wedge (u \wedge v)))))$
     $\wedge u = y \wedge (z \wedge (u \wedge v))$ [12,5]

20   $(((x \wedge y) \vee (z \vee y)) \wedge (u \vee (z \vee y))) \vee (v \wedge ((z \vee y) \wedge w)) = z \vee y$ [5,12]

21   $(((x \wedge (y \wedge (z \wedge u))) \vee z) \wedge (v \vee z)) \vee (y \wedge (z \wedge u)) = z$ [12,14,12,12]

22   $(((x \vee y) \wedge y) \vee (z \wedge y)) \wedge (u \vee (y \vee v)) = y$ [9,1]

23   $(((x \wedge y) \vee y) \wedge (z \wedge y)) \vee (u \wedge (v \wedge ((w \wedge y) \vee (y \wedge v6)))) = y$ [2,9,2,2,2]

24   $(((x \vee y) \wedge (z \wedge y)) \vee (u \wedge (z \wedge y))) \wedge (v \vee y) = z \wedge y$ [9,5]

25   $(((x \wedge y) \vee (z \vee y)) \wedge (u \vee (z \vee y))) \vee (v \wedge y) = z \vee y$ [5,9]

26   $(((x \wedge y) \vee y) \wedge (z \vee y)) \vee (u \wedge (v \wedge (y \wedge w))) = y$ [12,9,12,12,12]

27   $((x \vee (y \vee (x \vee z))) \wedge (u \vee (y \vee (x \vee z)))) \vee (x \vee z) = y \vee (x \vee z)$ [22,8]

28   $(((x \vee y) \wedge y) \vee (z \wedge y)) \wedge (u \vee (v \vee (y \vee w))) = y$ [22,5,22,22,22]

29   $(((x \vee (y \vee z)) \wedge (y \vee z)) \wedge (u \wedge z)) \wedge (v \vee ((y \vee z) \vee w)) = y \vee z$ [5,22]

30   $(((x \vee (((y \wedge z) \vee (u \vee z)) \wedge (v \vee (u \vee z)))) \wedge (((y \wedge z) \vee (u \vee z)) \wedge (v \vee$
     $(u \vee z)))) \vee (w \wedge (((y \wedge z) \vee (u \vee z)) \wedge (v \vee (u \vee z))))) \wedge (v6 \vee (u \vee z)) =$
     $((y \wedge z) \vee (u \vee z)) \wedge (v \vee (u \vee z))$ [14,22]

31   $(((x \vee (((y \wedge (z \wedge (u \wedge v))) \vee u) \wedge (w \vee u))) \wedge (((y \wedge (z \wedge (u \wedge v))) \vee u) \wedge (w$
     $\vee u))) \vee (v6 \wedge (((y \wedge (z \wedge (u \wedge v))) \vee u) \wedge (w \vee u)))) \wedge (v7 \vee u) =$

$((y \wedge (z \wedge (u \wedge v))) \vee u) \wedge (w \vee u)$ [21,22]

32 $(((x \wedge (y \vee (z \vee (u \vee v)))) \vee (y \vee (z \vee (u \vee v)))) \wedge (w \vee (y \vee (z \vee (u \vee v)))))$
$\vee u = y \vee (z \vee (u \vee v))$ [22,9]

33 $((((((x \vee y) \wedge y) \vee (z \wedge y)) \wedge (u \wedge y)) \vee (v \wedge (u \wedge y))) \wedge y = u \wedge y$ [1,17,1,1,1]

34 $(((x \wedge (y \wedge (z \wedge u))) \vee z) \wedge (v \vee z)) \wedge (w \wedge (z \wedge v6)) = z$ [12,20,12,12,12]

35 $(((x \vee y) \wedge y) \vee (z \wedge (u \wedge (y \wedge v)))) \wedge (w \vee ((v6 \vee y) \wedge (y \vee v7)))$
$= y$ [12,10,12,12,12,12]

36 $((((((x \wedge y) \vee y) \wedge (z \vee y)) \vee (u \vee y)) \wedge (v \vee (u \vee y))) \vee y = u \vee y$ [2,27,2,2,2]

37 $(((x \wedge (y \wedge (z \wedge u))) \vee z) \wedge (v \vee z)) \wedge (w \wedge ((v6 \wedge z) \vee (z \wedge v7)))$
$= z$ [12,11,12,12,12,12]

38 $(((x \wedge (y \wedge ((z \vee u) \wedge v))) \vee (z \vee u)) \wedge (w \vee (z \vee u))) \vee u = z \vee u$ [1,37]

39 $((((((x \vee y) \wedge y) \vee (z \wedge y)) \wedge (u \wedge y)) \vee (v \wedge (u \wedge y))) \wedge (w \vee y)$
$= u \wedge y$ [1,18,1,1,1]

40 $(((x \vee y) \wedge y) \vee (z \wedge y)) \wedge (((u \vee (y \vee v)) \wedge (y \vee v)) \vee (w \wedge (y \vee v))) = y$
[15,1]

41 $(((x \wedge y) \vee y) \wedge (z \vee y)) \wedge (u \wedge (v \wedge (((w \wedge y) \vee (y \wedge v6)) \wedge v7))) = y$ [19,2]

42 $(((x \wedge y) \vee (z \vee y)) \wedge (u \vee (z \vee y))) \vee (v \wedge (w \wedge (y \wedge v6))) = z \vee y$ [19,25]

43 $(((x \wedge y) \vee y) \wedge (z \vee y)) \vee (u \wedge (v \wedge y)) = y$ [1,41]

44 $(((x \wedge y) \vee y) \wedge (z \vee y)) \vee (u \wedge (v \wedge (w \wedge y))) = y$ [43,9,43,43,43]

45 $(((x \vee y) \wedge y) \vee (z \wedge (u \wedge (v \wedge y)))) \wedge (w \vee (y \vee v6)) = y$ [43,29,43,43,43]

46 $(((x \wedge (y \wedge (z \wedge u))) \vee z) \wedge (v \vee z)) \vee (w \wedge (v6 \wedge z)) = z$ [43,38,43,43,43]

47 $(((x \wedge y) \vee y) \wedge (z \vee y)) \vee (u \wedge (v \wedge ((w \wedge (v6 \wedge y)) \wedge v7))) = y$ [19,44]

48 $(((x \vee y) \wedge y) \vee (z \wedge (u \wedge ((v \wedge (w \wedge y)) \wedge v6)))) \wedge (v7 \vee (y \vee v8))$
$= y$ [19,45]

49 $(((x \vee y) \wedge (z \wedge y)) \vee (u \wedge (z \wedge y))) \wedge (v \vee (w \vee (y \vee v6))) = z \wedge y$ [32,24]

50 $(((x \wedge (y \vee z)) \vee (y \vee z)) \wedge (u \vee (y \vee z))) \vee y = y \vee z$ [40,2]

51 $(((x \vee y) \wedge y) \vee (z \wedge y)) \wedge (u \vee ((y \vee v) \wedge (y \vee w))) = y$ [40,6,40,40,40,40]

52 $(((x \wedge y) \vee (z \vee y)) \wedge (u \vee (z \vee y))) \vee z = z \vee y$ [40,11]

53 $(((x \wedge y) \vee y) \wedge (y \vee z)) \vee (u \wedge (y \wedge v)) = y$ [40,16,51,51,51]

54 $(((x \wedge y) \vee (z \vee y)) \wedge (u \vee (z \vee y))) \vee (v \wedge z) = z \vee y$ [53,11]

55 $(((x \vee y) \wedge y) \vee (z \wedge y)) \wedge (((u \vee y) \wedge (y \vee v)) \wedge w) = y$ [52,1]

56 $(((x \wedge (y \wedge (z \wedge u))) \vee z) \wedge (v \vee z)) \vee (((w \wedge (v6 \wedge (z \wedge v7))) \vee z) \wedge (v8 \vee z))$
$= z$ [34,52,34,34]

57 $(((x \wedge y) \vee (z \vee y)) \wedge (u \vee (z \vee y))) \vee (v \wedge (((w \wedge y) \vee (z \vee y)) \wedge (v6 \vee (z \vee y))$
$)))) = z \vee y$ [14,54,14,14]

58 $(((x \wedge (y \wedge (z \wedge u))) \vee z) \wedge (v \vee z)) \wedge (w \wedge (((v6 \wedge (v7 \wedge (z \wedge v8))) \vee z) \wedge$
$(v9 \vee z))) = z$ [34,54,34,34]

59 $(((x \vee y) \wedge y) \vee (z \wedge y)) \wedge y = y$ [2,55]

60 $((x \vee x) \wedge (y \vee x)) \vee (z \wedge (x \wedge u)) = x$ [59,12]

61 $((x \vee x) \wedge (y \vee x)) \vee (z \wedge (u \wedge (x \wedge v))) = x$ [59,26]

62 $((x \vee x) \wedge (y \vee x)) \vee (z \wedge (u \wedge ((v \wedge x) \vee (x \wedge w)))) = x$ [59,23]

63 $(x \vee (y \wedge x)) \wedge x = x$ [59,33,59,59,59]

64 $(((x \wedge (y \wedge (z \wedge u))) \vee z) \wedge (v \vee z)) \vee ((w \wedge z) \vee (z \wedge v6)) = z$ [59,37]

65 $(((x \wedge (y \wedge (z \wedge u))) \vee z) \wedge (v \vee z)) \vee (w \wedge z) = z$ [59,46]

66 $((x \vee x) \wedge (y \vee x)) \vee (z \wedge (u \wedge ((v \wedge (w \wedge x)) \wedge v6))) = x$ [59,47]

67    $(((x \wedge y) \vee (z \vee y)) \wedge (u \vee (z \vee y))) \vee (v \wedge (y \wedge w)) = z \vee y$ [59,42]
68    $(x \vee (y \wedge x)) \wedge (z \vee x) = x$ [59,39,59,59,59]
69    $(((x \vee y) \vee (x \vee y)) \wedge (z \vee (x \vee y))) \vee x = x \vee y$ [59,50]
70    $(x \vee (y \wedge (z \wedge ((u \wedge x) \vee (x \wedge v))))) \wedge x = x$ [13,63]
71    $(x \vee x) \wedge x = x$ [59,63]
72    $(x \vee (y \wedge (z \wedge ((u \wedge (v \wedge x)) \wedge w)))) \wedge (v6 \vee (x \vee v7)) = x$ [71,48]
73    $((x \wedge y) \vee (z \vee y)) \wedge (u \vee (z \vee y)) = (z \vee y) \wedge (v \vee (z \vee y))$ [71,30,57]
74    $((x \wedge (y \wedge (z \wedge u))) \vee z) \wedge (v \vee z) = z \wedge (w \vee z)$ [71,31,58]
75    $(x \vee (y \wedge (z \wedge ((u \wedge x) \vee (x \wedge v))))) \wedge (w \vee x) = x$ [13,68]
76    $((x \wedge (y \wedge z)) \vee (u \wedge (x \wedge (y \wedge z)))) \wedge z = x \wedge (y \wedge z)$ [43,68]
77    $(x \vee x) \wedge (y \vee x) = x$ [59,68]
78    $(x \vee y) \vee x = x \vee y$ [69,77]
79    $x \vee (y \wedge (z \wedge ((u \wedge (v \wedge x)) \wedge w))) = x$ [66,77]
80    $x \vee (y \wedge (z \wedge ((u \wedge x) \vee (x \wedge v)))) = x$ [62,77]
81    $x \vee (y \wedge (z \wedge (x \wedge u))) = x$ [61,77]
82    $x \vee (y \wedge (x \wedge z)) = x$ [60,77]
83    $x \wedge (y \vee (x \vee z)) = x$ [72,79]
84    $x \wedge (y \vee x) = x$ [75,80]
85    $x \wedge x = x$ [70,80]
86    $((B \wedge A) \vee (A \wedge C)) \vee A \neq A \;\mid\; ((B \vee A) \wedge (A \vee C)) \wedge A \neq A$ [3,82,83]
87    $((x \wedge y) \vee (z \vee y)) \wedge (u \vee (z \vee y)) = z \vee y$ [73,84]
88    $((x \wedge (y \wedge (z \wedge u))) \vee z) \wedge (v \vee z) = z$ [74,84]
89    $(x \vee y) \vee (z \wedge (y \wedge u)) = x \vee y$ [67,87]
90    $x \vee (y \wedge x) = x$ [65,88]
91    $x \vee ((y \wedge x) \vee (x \wedge z)) = x$ [64,88]
92    $x \vee x = x$ [56,88,88]
93    $(x \wedge (y \wedge z)) \wedge z = x \wedge (y \wedge z)$ [76,90]
94    $(x \wedge y) \vee y = y$ [85,12,89]
95    $x \vee (y \vee x) = y \vee x$ [85,36,94,84,78]
96    $(x \wedge y) \wedge (z \vee (u \vee (y \vee v))) = x \wedge y$ [85,49,94]
97    $(x \vee y) \wedge y = y$ [92,28,96]
98    $x \wedge (y \wedge x) = y \wedge x$ [92,33,97,90,93]
99    $x \wedge ((y \vee x) \wedge (x \vee z)) = x$ [92,35,97,81]
100    $((x \wedge y) \vee (y \wedge z)) \vee y = y$ [91,95,91]
101    $((B \vee A) \wedge (A \vee C)) \wedge A \neq A$ [86,100]
102    $((x \vee y) \wedge (y \vee z)) \wedge y = y$ [99,98,99]
103    $\square$ [102,101]

Huntington Identity, Proof of Order Reversibility

Here it is proved that if a uniquely complemented lattice satisfies axiom (8) in Theorem 4.8.4, then complementation is antitone. In view of

Theorem 4.8.1(1), this implies Theorem 4.8.4(8).

Input file:

```
formulas(assumptions).

  % Lattice Theory

  x v y = y v x.
  x ^ y = y ^ x.
  (x v y) v z = x v (y v z).
  (x ^ y) ^ z = x ^ (y ^ z).
  x ^ (x v y) = x.
  x v (x ^ y) = x.

  % Complementation

  x v x' = 1.
  x ^ x' = 0.

  % Unique Complementation

  x v y != 1 | x ^ y != 0 | x' = y # label(Unique_complementation).

  % Identity H82

  (x ^ y) v (x ^ z) = x ^ ((y ^ (x v z)) v (z ^ (x v y))) # label(H82).

  % Denial of order reversibility

  A ^ B = A.
  A' v B' != A' # answer(Order_reversibility).

end_of_list.
```

Prover9 proof in 3 seconds:

1	$x \lor y = y \lor x$	[assumption]
2	$x \land y = y \land x$	[assumption]
3	$(x \lor y) \lor z = x \lor (y \lor z)$	[assumption]
4	$(x \land y) \land z = x \land (y \land z)$	[assumption]
5	$x \land (x \lor y) = x$	[assumption]
6	$x \lor (x \land y) = x$	[assumption]
7	$x \lor x' = 1$	[assumption]
8	$x \land x' = 0$	[assumption]

9	$x \lor y \neq 1 \mid x \land y \neq 0 \mid x' = y$ # Unique_complementation	[assumption]
10	$(x \land y) \lor (x \land z) = x \land ((y \land (x \lor z)) \lor (z \land (x \lor y)))$ # H82	[assumption]
11	$A \land B = A$	[assumption]
12	$A' \lor B' \neq A'$ # Order_reversibility	[assumption]
13	$x \land (y \land z) = y \land (x \land z)$	[2,4,4]
14	$x \land (y \lor x) = x$	[1,5]
15	$x \lor ((x \land y) \lor z) = x \lor z$	[6,3]
16	$x \lor (x' \lor y) = 1 \lor y$	[7,3]
17	$x \land 1 = x$	[7,5]
18	$x \lor 0 = x$	[8,6]
19	$0 \lor (x \land y) = x \land (y \lor (x' \land (x \lor y)))$	[7,10,8,1,17]
20	$A \land (B \land x) = A \land x$	[11,4]
21	$1 \land x = x$	[17,2]
22	$0 \lor x = x$	[18,1]
23	$x \land (y \lor (x' \land (x \lor y))) = x \land y$	[19,22]
24	$x \land (y \land x') = y \land 0$	[8,13]
25	$1 \lor x = 1$	[21,5]
26	$x \lor (x' \lor y) = 1$	[16,25]
27	$0 \land x = 0$	[22,5]
28	$x \lor 1 = 1$	[14,21]
29	$x \land 0 = 0$	[27,2]
30	$x \land (y \land x') = 0$	[24,29]
31	$A \land (A' \lor B') \neq 0$ # Order_reversibility	[9,26,12]
32	$x \lor (x \land y)' = 1$	[7,15,28]
33	$x \lor (y \land x)' = 1$	[2,32]
34	$B \lor A' = 1$	[11,33]
35	$B \land (A' \lor B') = B \land A'$	[34,23,2,21]
36	$A \land (A' \lor B') = 0$	[35,20,30]
37	$\Box$ # Order_reversibility	[36,31]

Huntington Identity, Proof of Distributivity

Here we prove Theorem 4.8.4(8).

Input file:

```
formulas(assumptions).

% Lattice Theory

x v y = y v x.
x ^ y = y ^ x.
```

```
(x v y) v z = x v (y v z).
(x ^ y) ^ z = x ^ (y ^ z).
x ^ (x v y) = x.
x v (x ^ y) = x.
```

% Complementation

```
x v x' = 1.
x ^ x' = 0.
```

% Unique Complementation

```
x v y != 1 | x ^ y != 0 | x' = y # label(Unique_complementation).
```

% Identity H82

```
(x ^ y) v (x ^ z) = x ^ ((y ^ (x v z)) v (z ^ (x v y))) # label(H82).
```

% Denial of distributivity

```
(A ^ B) v (A ^ C) != A ^ (B v C)  # answer(Distributivity).
```

```
end_of_list.
```

Prover9 proof in 85 seconds:

1	$x \vee y = y \vee x$	[assumption]
2	$x \wedge y = y \wedge x$	[assumption]
3	$(x \vee y) \vee z = x \vee (y \vee z)$	[assumption]
4	$(x \wedge y) \wedge z = x \wedge (y \wedge z)$	[assumption]
5	$x \wedge (x \vee y) = x$	[assumption]
6	$x \vee (x \wedge y) = x$	[assumption]
7	$x \vee x' = 1$	[assumption]
8	$x \wedge x' = 0$	[assumption]
9	$x \vee y \neq 1 \mid x \wedge y \neq 0 \mid x' = y$ # Unique_complementation	[assumption]
10	$(x \wedge y) \vee (x \wedge z) = x \wedge ((y \wedge (x \vee z)) \vee (z \wedge (x \vee y)))$ # H82	[assumption]
11	$(A \wedge B) \vee (A \wedge C) \neq A \wedge (B \vee C)$ # Distributivity	[assumption]
12	$x \vee (y \vee z) = y \vee (x \vee z)$	[1,3,3]
13	$x \wedge (y \wedge z) = y \wedge (x \wedge z)$	[2,4,4]
14	$x \wedge (y \vee x) = x$	[1,5]
15	$x \wedge ((x \vee y) \wedge z) = x \wedge z$	[5,4]
16	$x \vee (y \wedge x) = x$	[2,6]
17	$x \vee ((x \wedge y) \vee z) = x \vee z$	[6,3]
18	$(x \wedge y) \vee (x \wedge (y \wedge z)) = x \wedge y$	[4,6]
19	$x \wedge x = x$	[6,5]

20	$x \vee (x' \vee y) = 1 \vee y$	[7,3]
21	$x \vee (y \vee (x \vee y)') = 1$	[7,3]
22	$x \wedge 1 = x$	[7,5]
23	$x \wedge (y \wedge (x \wedge y)') = 0$	[8,4]
24	$x \vee 0 = x$	[8,6]
25	$x \vee y \neq 1 \mid y \wedge x \neq 0 \mid y' = x$	[1,9]
26	$(x \wedge y) \vee z \neq 1 \mid x \wedge (y \wedge z) \neq 0 \mid (x \wedge y)' = z$	[4,9]
27	$x \wedge ((y \wedge (x \vee z)) \vee (z \wedge (x \vee y))) = (x \wedge z) \vee (x \wedge y)$	[10,1]
28	$0 \vee (x \wedge y) = x \wedge (y \vee (x' \wedge (x \vee y)))$	[7,10,8,1,22]
29	$x \wedge (y \wedge x) = y \wedge x$	[19,4,2]
30	$1 \wedge x = x$	[22,2]
31	$0 \vee x = x$	[24,1]
32	$x \wedge (y \vee (x' \wedge (x \vee y))) = x \wedge y$	[28,31]
33	$x \wedge (y \wedge (x \vee z)) = y \wedge x$	[5,13]
34	$1 \vee x = 1$	[30,5]
35	$x \vee (x' \vee y) = 1$	[20,34]
36	$0 \wedge x = 0$	[31,5]
37	$x \wedge ((y \vee x) \wedge z) = x \wedge z$	[14,4]
38	$x \wedge (y \wedge (z \vee x)) = y \wedge x$	[14,13]
39	$x \vee 1 = 1$	[14,30]
40	$x \wedge 0 = 0$	[36,2]
41	$x \vee ((y \wedge x) \vee z) = x \vee z$	[16,3]
42	$x \wedge (x \vee y)' = 0$	[8,15,40]
43	$x \wedge (y \vee x)' = 0$	[1,42]
44	$x \wedge (y \vee (z \vee x))' = 0$	[3,43]
45	$x \wedge (y \wedge (z \vee (x \wedge y))') = 0$	[43,4]
46	$x \vee (x \wedge y)' = 1$	[7,17,39]
47	$x \vee (y \wedge x)' = 1$	[2,46]
48	$x \vee (y \vee ((x \vee y) \wedge z)') = 1$	[46,3]
49	$(x \wedge y) \vee (y \wedge x) = x \wedge y$	[29,18]
50	$x \wedge (y \wedge (y \wedge x)') = 0$	[2,23]
51	$x \vee y \neq 1 \mid x \wedge y \neq 0 \mid y' = x$	[2,25]
52	$x'' = x$	[7,25,2,8]
53	$x \wedge (y \wedge x)' \neq 0 \mid y \wedge x = x$	[47,25,2,52]
54	$x \wedge (y \wedge ((x \wedge y)' \vee z)) \neq 0 \mid (x \wedge y)' \vee z = (x \wedge y)'$	[35,26]
55	$x \wedge ((y \wedge (z \vee x)) \vee (z \wedge (x \vee y))) = (x \wedge z) \vee (x \wedge y)$	[1,27]
56	$x \wedge (y \wedge (y \wedge (x \vee z))') = 0$	[50,15,40]
57	$x \wedge (y \wedge (z \vee (y \wedge x))') = 0$	[49,44,4]
58	$x \wedge (x' \vee (y \vee (x \vee y)')) = x \wedge (y \vee (x \vee y)')$	[21,32,2,30,1]
59	$x \wedge ((y \vee x) \wedge (x \vee z))' = 0$	[56,37]
60	$(x \vee y) \wedge x' \neq 0 \mid x \vee y = x$	[5,53,5]
61	$(x \vee y) \wedge y' \neq 0 \mid y \vee x = y$	[1,60]
62	$x' \vee (y \wedge x)' = (y \wedge x)'$	[32,54,23,47,2,30,1]
63	$x' \wedge (y \wedge x)' = x'$	[62,5]
64	$x' \vee (x \vee y)' = x'$	[5,62,1,5]

65	$x \vee (y \wedge (x \vee y'))' = 1$	[62,48,2]
66	$x' \wedge (x \vee y)' = (x \vee y)'$	[5,63,2]
67	$x' \vee (y \vee (x \vee z)') = y \vee x'$	[64,12]
68	$x \wedge (y \vee (x \vee y)') = x \wedge (y \vee x')$	[58,67]
69	$(x \vee y) \wedge (y \vee (x \vee y)') = y$	[51,65,59,52]
70	$x \vee (y' \wedge (x \vee y))' = 1$	[52,65]
71	$x \vee ((x \vee y) \wedge y')' = 1$	[2,70]
72	$x \wedge (y \vee (x \wedge y'))' = 0$	[66,45]
73	$x \wedge (y \vee (y' \wedge x))' = 0$	[66,57]
74	$x \wedge (y' \vee (x \wedge y))' = 0$	[52,72]
75	$x \wedge (y' \vee (y \wedge x))' = 0$	[52,73]
76	$x \wedge ((x \wedge y) \vee y')' = 0$	[1,74]
77	$x \wedge ((y \wedge x) \vee y')' = 0$	[1,75]
78	$x \wedge (y \vee x') = x \wedge y$	[69,15,68]
79	$x \wedge (y \vee (z \vee x')) = x \wedge (y \vee z)$	[3,78]
80	$(x \vee y) \wedge y' = x \wedge y'$	[71,78,2,30,2,15]
81	$x \wedge y' \neq 0 \mid y \vee x = y$	[61,80]
82	$(x \wedge y) \vee x' = y \vee x'$	[81,77,1,41]
83	$(x \wedge y) \vee y' = x \vee y'$	[81,76,1,17]
84	$(x \wedge (y \vee z)) \vee y' = x \vee y'$	[33,82,83]
85	$(x \wedge (y \vee z)) \vee z' = x \vee z'$	[38,82,83]
86	$(x \wedge y) \vee (x \wedge z) = x \wedge (y \vee z)$	[55,78,3,84,12,85,79]
87	$\square$ # Distributivity	[86,11]

References

[1] W. McCune. Prover9. http://www.cs.unm.edu/~mccune/prover9/, 2005–2007.

Appendix B: Partially Ordered Sets and Betweenness

We have not found the items marked by *, which we quote from Bull. Sci. Math. 15(1891), 53-68; Jahr. Fortschr. Math. 35(1904), 82; 53(1927), 44, Huntington [1917], [1935] Yooneyama [1918] and Katriňák [1959].

It is well known that with every relation of *partial order*, i.e., a binary relation $\leq$ which satisfies

$$(1) \qquad x \leq x \,,$$

$$(2) \qquad x \leq y \ \& \ y \leq x \Longrightarrow x = y \,,$$

$$(3) \qquad x \leq y \ \& \ y \leq z \Longrightarrow x \leq z \,,$$

is associated the relation $<$ defined by

$$(4) \qquad x < y \Longleftrightarrow x \leq y \ \& \ x \neq y \,.$$

This relation satisfies

$$(5) \qquad x \not< x \,,$$

$$(6) \qquad x < y \Longrightarrow y \not< x \,,$$

$$(7) \qquad x < y \ \& \ y < z \Longrightarrow x < z \,;$$

note that (6) implies (5).
Conversely, with any relation $<$ of *strict partial order*, i.e., satisfying (5)–(7), is associated a partial order $\leq$ defined by

$$(8) \qquad x \leq y \Longleftrightarrow x < y \text{ or } x = y \,.$$

The correspondence (4), (8) is one-to-one. So

$$\mathbf{P}_1 = \{(1), (2), (3)\}$$

and

$$\mathbf{P}_2 = \{(6), (7)\}$$

are independent sets of axioms for *partially ordered sets* or *posets*, i.e., sets endowed with a partial order.

The founders of the theory of partially ordered sets are Peirce [1880,87], Schröder [1890-1905] and Hausdorff [1914]; cf. Birkhoff [1948], Ch.I. footnote 1.   As a matter of fact, in those early days the main interest was concentrated on what we call today *totally ordered sets* or *chains*, that is, sets endowed with a relation $<$ which satisfies (5)–(7) and

(9) $$x \neq y \Longrightarrow x < y \text{ or } y < x .$$

Besides, in the years 1890-1930 the abstract relation $<$ was not yet well crystallized and many authors worked with $<$, $>$ and $=$ as three interconnected primitive concepts. So did Betazzi [1890]*, Burali-Forti [1893]*, Stoltz and Gmeiner [1901]*, Shatunovskiĭ [1904]*, Yooneyama [1918] and Su [1927]*.

The first axiom systems for totally ordered sets use the above axioms and

(6′) $$\neg(x < y \,\&\, y < x) ,$$

(10) $$x < y \Longrightarrow x \neq y ,$$

(11) $$\neg(x \neq y \,\&\, x < y \,\&\, y < x) ,$$

(12) $$x, y, z \text{ distinct } \&\, x < y \,\&\, y < z \Longrightarrow x < z ,$$

(13) $$x < y \Longrightarrow z < y \text{ or } x < z .$$

The very first system seems to be

$$\mathbf{T}_1 = \{(6'), (7), (9)\} ,$$

given by Vailatti [1892]*, while Huntington [1904-6], [1917] found the systems

$$\mathbf{T}_2 = \{(6), (7), (9)\} ,$$

$$\mathbf{T}_3 = \{(7), (9), (10)\} ,$$

$$\mathbf{T}_4 = \{(5), (9), (11), (12)\} ,$$

$$\mathbf{T}_5 = \{(6'), (9), (12)\} \,,$$

and proved that they are completely independent. Later on, Huntington [1935] devised the system

$$\mathbf{T}_6 = \{(5), (9), (11), (13)\} \,.$$

So $\mathbf{T}_1 - \mathbf{T}_5$ are maximal independent subsystems of $\{(1), \ldots, (12)\}$, hence they are logically equivalent to it, but it is not known whether there are also other maximal independent subsystems. The complete existential theory of system $\{(1), \ldots, (12)\}$ was not studied.

Chains have been also characterized in terms of the relation of *open betweenness*, defined by

$$(14) \qquad\qquad (xyz) \Longleftrightarrow x < y < z \text{ or } z < y < x \,.$$

It is easy to check the following properties:

$$(15) \qquad\qquad x, y, z \text{ distinct} \,\&\, (xyz) \Longrightarrow (zyx) \,,$$

$$(16) \quad x, y, z \text{ distinct} \Longrightarrow (xyz) \text{ or } (xzy) \text{ or } (yxz) \text{ or } (yzx) \text{ or } (xzy) \text{ or } (zyx) \,,$$

$$(17) \qquad\qquad x, y, z \text{ distinct} \Longrightarrow \neg((xyz) \,\&\, (xzy)) \,,$$

$$(18) \qquad\qquad (xyz) \Longrightarrow x, y, z \text{ distinct} \,,$$

$$(19) \qquad\qquad x, y, z, t \text{ distinct} \,\&\, (xyz) \,\&\, (yzt) \Longrightarrow (xyt) \,,$$

$$(20) \qquad\qquad x, y, z, t \text{ distinct} \,\&\, (xyz) \,\&\, (ytz) \Longrightarrow (xyt) \,,$$

$$(21) \qquad\qquad x, y, z, t \text{ distinct} \,\&\, (xyz) \,\&\, (ytz) \Longrightarrow (xtz) \,,$$

$$(22) \qquad\qquad x, y, z, t \text{ distinct} \,\&\, (xyz) \,\&\, (xtz) \Longrightarrow (xyt) \text{ or } (xty) \,,$$

$$(23) \qquad\qquad x, y, z, t \text{ distinct} \,\&\, (xyz) \,\&\, (xtz) \Longrightarrow (xyt) \text{ or } (tyz) \,,$$

$$(24) \qquad\qquad x, y, z, t \text{ distinct} \,\&\, (xyz) \,\&\, (tyz) \Longrightarrow (xtz) \text{ or } (txz) \,,$$

$$(25) \qquad\qquad x, y, z, t \text{ distinct} \,\&\, (xyz) \,\&\, (tyz) \Longrightarrow (xty) \text{ or } (txy) \,,$$

$$(26) \qquad\qquad x, y, z, t \text{ distinct} \,\&\, (xyz) \,\&\, (tyz) \Longrightarrow (xty) \text{ or } (txz) \,,$$

$$(27) \qquad\qquad x, y, z, t \text{ distinct} \,\&\, (xyz) \Longrightarrow (xyt) \text{ or } (tyz) \,.$$

Huntington and Kline [1917] and Huntington [1924a] have proved that the following axiom systems define totally ordered sets and are independent:

$$\mathbf{T}_7 = \{(15), (16), (17), (18), (19), (20)\},$$

$$\mathbf{T}_8 = \{(15), (16), (17), (18), (19), (23)\},$$

$$\mathbf{T}_9 = \{(15), (16), (17), (18), (19), (24)\},$$

$$\mathbf{T}_{10} = \{(15), (16), (17), (18), (19), (25)\},$$

$$\mathbf{T}_{11} = \{(15), (16), (17), (18), (19), (26)\},$$

$$\mathbf{T}_{12} = \{(15), (16), (17), (18), (20), (22)\},$$

$$\mathbf{T}_{13} = \{(15), (16), (17), (18), (20), (23)\},$$

$$\mathbf{T}_{14} = \{(15), (16), (17), (18), (21), (23)\},$$

$$\mathbf{T}_{15} = \{(15), (16), (17), (18), (21), (22), (24)\},$$

$$\mathbf{T}_{16} = \{(15), (16), (17), (18), (21), (22), (25)\},$$

$$\mathbf{T}_{17} = \{(15), (16), (17), (18), (21), (22), (26)\},$$

$$\mathbf{T}_{18} = \{(15), (16), (17), (18), (27)\},$$

among which $\mathbf{T}_{18}$ is completely independent. Van de Walle [1924] proved that systems $\{\mathbf{T}_7, \ldots, \mathbf{T}_{16}\}$ are also completely independent, while $\mathbf{T}_{17}$ does not share this property. Rosenbaum [1951] gave the following completely independent axiom system for chains:

$$\mathbf{T}_{19} = \{(15), (16), (18), (28), (29)\},$$

where the last two axioms are weaker variants of (19) and (20), respectively:

(28) $$(xyz) \ \& \ (yzt) \Longrightarrow (xyt),$$

(29) $$(xyz) \ \& \ (ytz) \Longrightarrow (xyt).$$

The characterizations of chains by each system $\mathbf{T}_i$, $(i = 7, \ldots, 19)$ has the following meaning. The total order $<$ satisfies $\mathbf{T}_i$ and conversely, if a ternary relation on a set T satisfies $\mathbf{T}_i$, then for each pair of distinct elements $a, b \in T$, the following relation $<_{a,b}$ is a total order on T:

(30.1) $a <_{a,b} b \ \& \ \neg(b <_{a,b} a) \ \& \ \neg(x <_{a,b} x)$,

(30.2) $x <_{a,b} a \Longleftrightarrow (xab)$,

(30.3) $x <_{a,b} b \Longleftrightarrow (xab)$ or (abx) ,

(30.4) $a <_{a,b} x \Longleftrightarrow (axb)$ or (abx) ,

(30.5) $b <_{a,b} x \Longleftrightarrow (abx)$,

(30.6)

$x <_{a,b} y \Longleftrightarrow (xab) \ \& \ (xyb)$ or $(xab) \ \& \ (xby)$ or

or $(axb) \ \& \ (axy)$ or $(abx) \ \& \ (axy)$.

Moreover, the betweenness relation (14) associated with $<_{a,b}$ coincides with the original ternary relation. On the other hand, if the starting point is a chain $(T, <)$ and $a < b$ in T, then the construction (30) applied to the betweenness relation (14) yields $<_{a,b} \, = \, <$.

Another system, whose independence has not been studied, was given by Moisil [1942], namely

$$\mathbf{T}_{20} = \{(15), 19), (20), (31), (32)\} \, ,$$

where we have set

(31) $\neg(xxy) \ \& \ \neg(xyx)$,

(32) x, y, z distinct $\Longrightarrow (xyz)$ or (yzx) or (zxy) .

Čech [1936]* has characterized totally ordered sets with at least three elements by the system

$$\mathbf{T}_{21} = \{(32), (33), (34), (35), (36), (37), (38), (39)\} \, ,$$

where we have set

(33) $(xyz) \Longrightarrow x \neq z$,

(34) $(xyz) \Longrightarrow x \neq y$,

(35) $(xyz) \Longrightarrow (zyx)$,

(36) $(xyz) \ \& \ (yzt) \Longrightarrow (xzt)$,

(37) $(xyz) \ \& \ (xzt) \Longrightarrow (xyt)$,

(38) $(xyz) \ \& \ (xtz) \ \& \ y \neq t \Longrightarrow (xyt)$ or (xty) ,

(39) (xyz) & (xyt) & $z \neq t \Longrightarrow (yzt)$ or (ytz) .

For every set of cardinality at least 3 and endowed with a ternary relation which satisfies system $\mathbf{T}_{21}$, there are exactly two relations of total order such that the betweenness relation (14) associated with them coincides with the original ternary relation (these two relations are dual to each other). This property is "better" than the corresponding property of the total orders $<_{a,b}$.

Katriňák [1959] has shown that system $\mathbf{T}_{21}$ is not independent, while systems

$$\mathbf{T}_{22} = \{(32), (34), (35), (36), (37), (39)\} ,$$

$$\mathbf{T}_{23} = \{(32), (33), (35), (36), (39)\} ,$$

$$\mathbf{T}_{24} = \{(32), (34), (35), (37), (38), (39)\} ,$$

are independent and logically equivalent to $\mathbf{T}_{21}$.

Shepperd [1956] has characterized totally ordered sets in termes of the relation of *closed betweenness*, defined by

(40) $xyz \Longleftrightarrow x \leq y \leq z$ or $z \leq y \leq x$.

This is a specialization of Smiley and Transue's relation of lattice betweennes, which can be used to define lattices; cf. Chapter 1, §4, (1.4.12). The axiom system provided by Shepperd is

$$\mathbf{T}_{25} = \{(41), (42), (43), (44), (45)\} ,$$

where we have set

(41) xyz or yzx or zxy ,

(42) xyz & $xzy \Longrightarrow y = z$,

(43) $xyz \Longrightarrow zyx$,

(44) xyz & $xzt \Longrightarrow yzt$,

(45) xyz & yzt & $y \neq z \Longrightarrow xyt$.

McPhee [1962] proved that systems

$$\mathbf{T}_{26} = \{(46), (47), (48), (49)\} \,,$$

$$\mathbf{T}_{27} = \{(48), (49), (50)\} \,,$$

$$\mathbf{T}_{28} = \{(47), (51), (52)\} \,,$$

are independent and logically equivalent to $\mathbf{T}_{25}$, where

(46) $\exists\, a\ xya$ or xay or yax or yxa or axy or ayx ,

(47) yxz & $ztx \implies xty$,

(48) yxz & $tyx \implies zxt$ or $x = y$,

(49) yxz & zxt & $txy \implies x = y$ or $x = z$ or $x = t$,

(50) xyz or yzx or yxz ,

(51) xyz or xzy or yzx or yxz or zxy or zyx ,

(52) zxt & zyt & xzy & $xty \implies x = z$ or $y = z$ or $x = t$ or $y = t$.

A ternary relation which appears in geometry is the cyclic order $\circ\, ABC$ between the points of an oriented circle. This relation means that the arc ABC is positively oriented.

More generally, the relation of *cyclic order* is defined on a totally ordered set by

(53) $\circ\, xyz \iff x < y < z$ or $y < z < x$ or $z < x < y$.

The following properties are easily checked:

(54) x, y, z distinct & $\circ\, xyz \implies \circ\, yzx$,

(55) x, y, z distinct $\implies \circ\, xyz$ or $\circ\, yzx$ or $\circ\, zxy$ or $\circ\, zyx$ or $\circ\, yxz$ or $\circ\, xzy$,

(56) x, y, z distinct $\implies \neg(\circ\, xyz$ & $\circ\, xzy)$,

(57) $\circ\, xyz \implies x, y, z$ distinct ,

(58) x, y, z, t distinct & $\circ\, xyz$ & $\circ\, ytz \implies \circ\, xyt$,

(59) x, y, z, t distinct & $\circ\, xyz$ & $\circ\, ytz \implies \circ\, xtz$,

(60) x, y, z, t distinct & $\circ\, xyz \Longrightarrow \circ\, xyt$ or $\circ\, tyz$.

Huntington [1924b] has proved that the following axiom systems define totally ordered sets and are completely independent:

$$\mathbf{T}_{29} = \{(54), (55), (56), (57), (58)\} ,$$

$$\mathbf{T}_{30} = \{(54), (55), (56), (57), (59)\} ,$$

$$\mathbf{T}_{31} = \{(54), (55), (56), (57), (60)\} ,$$

and if the support set T has at least three elements, the system

$$\mathbf{T}_{32} = \{(54), (56), (57), (60), (61)\} ,$$

where axiom (61) states that there exist distinct elements a, b, c satisfying $\circ\, abc$, defines also chains.

The characterization of chains by each system $\mathbf{T}_i$ $(i = 29, \ldots, 32)$ has the following meaning. The total order $<$ satisfies $\mathbf{T}_i$ and conversely, if a ternary relation on a set T satisfies $\mathbf{T}_i$, then for each element $a \in T$ the following relation $<_a$ is a total order on T having a as least element:

(62.1) $x <_a y \Longleftrightarrow \circ\, axy$,

(62.2) $\neg(x <_a x)$,

(62.3) $x \neq a \Longrightarrow a <_a x \,\&\, \neg(x <_a a)$.

Moreover, the following two properties I and II hold:

I. The relation of cyclic order (53) associated with $<_a$ coincides with the original ternary relation.

Proof. If the elements $x, y, z \in T$ satisfy relation (53) corresponding to $<_a$, then they are distinct and by (62.3) they are also distinct from a. So it follows by (54) and (58) that each of the double inequalities $x <_a y <_a z$, $y <_a z <_a x$ and $z <_a x <_a y$ implies $\circ\, xyz$. Conversely, suppose $\circ\, xyz$ holds. Then x, y, z are distinct by (57). If one of these elements is a, in view of (54) we can suppose without loss of generality that $x = a$, hence $x = a <_a y <_a z$ by (62), therefore the right-hand side of (53) holds, as desired. Now suppose x, y, z are also distinct from a. Then (60) implies $\circ\, xya$ or $\circ\, ayz$; in the former case $\circ\, axy$, again by (54). So $x <_a y$ or $y <_a z$. Since $<_a$ is a total order, there are three possibilities: 1) $x <_a y <_a z$, or 2) $x <_a y$ and $z <_a y$, or 3) $y <_a x$ and $y <_a z$. The desired conclusion is already established in case 1). Case 2) yields the subcases $x <_a z <_a y$

and $z <_a x <_a y$; in each subcase the desired conclusion holds. Case 3) is treated similarly.

II. If the starting point is a chain $(T, <)$ having a as least element, then the construction (62) applied to the original cyclic order (53) yields $<_a = <$.

Proof. It follows by (62.1) and (53) that

$$y <_a x \Longleftrightarrow \circ ayz \Longleftrightarrow a < y < z \Longleftrightarrow y < z.$$

Another system, whose independence was not studied, was given by Moisil [1942], namely

$$\mathbf{T}_{33} = \{(54), (63), (64), (65)\},$$

where

(63) $$x, y, z \text{ distinct} \Longrightarrow \circ xyz \text{ or } \circ xzy,$$

(64) $$\neg \circ xxy,$$

(65) $$x, y, z \text{ distinct } \& \circ xyz \& \circ xzt \Longrightarrow xyt.$$

Another relation which appears in geometry is the separation relation $ABCD$ between the points of a circle, defined as follows: "the points AC separate the points BD", or equivalently, "the sum of the unoriented arcs ABC and ADC is 2π".

More generally, the *separation* relation is defined on a totally ordered set by

(66)
$$xyzt \Longleftrightarrow x < y < z < t \text{ or } y < z < t < x \text{ or } z < t < x < y \text{ or}$$
$$\text{or } t < x < y < z \text{ or } t < z < y < x \text{ or } x < t < z < y \text{ or}$$
$$\text{or } y < x < t < z \text{ or } z < y < x < t.$$

The following properties are satisfied:

(67) $$xyzt \Longrightarrow x, y, z, t \text{ distinct},$$

(68) $$x, y, z, t \text{ distinct } \& xyzt \Longrightarrow yztx,$$

(69) $$x, y, z, t \text{ distinct} \Longrightarrow \neg(xyzt \& xytz),$$

(70) $$x, y, z, t \text{ distinct } \& xyzt \Longrightarrow \exists a, b, c, d \text{ distinct } \& abcd \& dcba,$$

(71) $$x, y, z, t, u \text{ distinct } \& xyzt \Longrightarrow xuzt \text{ or } xyzu,$$

(72) $$x, y, z, t \text{ distinct} \Longrightarrow xyzt \text{ or } xytz \text{ or} \ldots \text{ or } tzyx,$$

(73) x, y, z, t distinct & $xyzt \implies tzyx$,

(74) x, y, z, t, u distinct & $xyzt$ & $xytu \implies xyzu$,

(75) x, y, z, t, u distinct & $xyzt$ & $xytu \implies yztu$,

(76) x, y, z, t, u distinct & $xyzt$ & $xytu \implies xztu$,

(77) x, y, z, t, u distinct & $xyzt$ & $xytu \implies xyzu$ or $xyuz$,

(78) x, y, z, t, u distinct & $xyzt$ & $xyzu \implies xztu$ or $xzut$,

(79) x, y, z, t, u distinct & $xyzt$ & $xyzu \implies yztu$ or $yzut$,

(80) x, y, z, t, u distinct & $xyzt$ & $xyzu \implies (xytu$ or $xzut)$ & $(xyut$ or $xztu)$,

(81) x, y, z, t, u distinct & $xyzt$ & $xyzu \implies (xytu$ or $yztu)$ & $(xyut$ or $yztu)$,

(82) z, y, z, t, u distinct & $xyzt$ & $xyzu \implies (xztu$ or $yzut)$ & $(xzut$ or $yztu)$.

Huntington and Rosinger [1932]* have proved that the following axiom systems define totally ordered sets with least element and are independent:

$$\mathbf{T}_{34} = \{(67), (68), (69), (71), (72), (73)\} ,$$

$$\mathbf{T}_{35} = \{(67), (68), (69), (72), (73), (75)\} ,$$

$$\mathbf{T}_{36} = \{(67), (68), (69), (72), (73), (76)\} ,$$

$$\mathbf{T}_{37} = \{(67), (68), (69), (72), (73), (74), (77)\} ,$$

$$\mathbf{T}_{38} = \{(67), (68), (69), (72), (73), (74), (78)\} ,$$

$$\mathbf{T}_{39} = \{(67), (68), (69), (72), (73), (74), (79)\} ,$$

$$\mathbf{T}_{40} = \{(67), (68), (69), (72), (73), (74), (80)\} ,$$

$$\mathbf{T}_{41} = \{(67), (68), (69), (72), (73), (74), (81)\} ,$$

$$\mathbf{T}_{42} = \{(67), (68), (69), (72), (73), (74), (82)\} ,$$

and if the support T has at least 4 elements, the system

$$\mathbf{T}_{43} = \{(67), (68), (70), (83)\} ,$$

where axiom (83) states that there exist distinct elements a, b, c, d satisfying $abcd$, defines also chains.

The characterization of chains by each system $\mathbf{T}_i$ ($i = 23, \ldots, 43$) has the following meaning. The total order $<$ satisfies $\mathbf{T}_i$ and conversely, if a quaternary relation on a set T satisfies $\mathbf{T}_i$, then for each triple of distinct elements $a, b, c \in T$ the following relation $<_{a,b,c}$ is a total order on T having a as least element:

(84.1) $$ b <_{a,b,c} c \,\&\, \neg(c <_{a,b,c} b) \,\&\, \neg(x <_{a,b,c} x) \,, $$

(84.2) $$ x \neq a \Longrightarrow a <_{a,b,c} x \,\&\, \neg(x <_{a,b,c} a) \,, $$

(84.3) $$ x <_{a,b,c} b \Longleftrightarrow axbc \,, $$

(84.4) $$ b <_{a,b,c} x \Longleftrightarrow abxc \text{ or } abcx \,, $$

(84.5) $$ x <_{a,b,c} c \Longleftrightarrow axbc \text{ or } abxc \,, $$

(84.6) $$ c <_{a,b,c} x \Longleftrightarrow abcx \,, $$

(84.7) $$ x <_{a,b,c} y \Longleftrightarrow (axbc \,\&\, axyb) \text{ or } (axbc \,\&\, axcy) \text{ or } $$
$$ \text{or } (abxc \,\&\, abxy) \text{ or } (abcx \,\&\, abxy) \,. $$

On the other hand, if the starting point is a chain $(T, <)$ having a least element A and if $b, c \in T$ satisfy $a < b < c$, then the construction (84) applied to the separation relation (66) yields $<_{a,b,c} = <$.

Shepperd [1956] has characterized totally ordered sets in terms of the relation of *closed separation*, defined by

(85) $$ \overline{xyzt} \Longleftrightarrow x \leq y \leq z \leq t \text{ or } y \leq z \leq t \leq x \text{ or } z \leq t \leq x \leq y \text{ or} $$
$$ \text{or } t \leq x \leq y \leq z \text{ or } t \leq z \leq y \leq x \text{ or} $$
$$ \text{or } x \leq t \leq z \leq y \text{ or } y \leq x \leq t \leq z \text{ or } z \leq y \leq x \leq t \,. $$

His axiom system is

$$ \mathbf{T}_{44} = \{(86), (87), (88), (89), (90)\} \,, $$

where

(86) $$ \overline{xyzt} \text{ or } \overline{xzyt} \text{ or } \overline{xytz} \,, $$

(87) $$ \overline{xyzt} \text{ or } \overline{xytz} \Longleftrightarrow x = y \text{ or } z = t \,, $$

(88) $\overline{xyzt} \Longrightarrow \overline{txyz}$,

(89) $\overline{xyzt} \Longrightarrow \overline{yztx}$,

(90) $\overline{xyzt}$ & $\overline{xuyz}$ & $y \neq z \Longrightarrow \overline{xuyt}$.

See systems $\mathbf{T}_{45}$-$\mathbf{T}_{50}$ after $\mathbf{P}_{13}$.

Problem 1 in Birkhoff [1948] asks for a characterization of partially ordered sets in terms of the closed betweenness relation

(40) $xyz \Longleftrightarrow x \leq y \leq z$ or $z \leq y \leq z$.

The first answer to this problem was given by Altwegg [1950], who devised the system

$$\mathbf{P}_3 = \{(43), (91), (92), (93), (94), (95), (96)\} \,,$$

where we have set

(91) xxx ,

(92) $xyz \Longrightarrow xxy$,

(93) $xyx \Longrightarrow x = y$,

(94) xyz & yzu & $y \neq z \Longrightarrow xyu$,

(95) x, y, z, t distinct & xyz & $ytt \Longrightarrow (xyt$ & $\neg tyz)$ or $(\neg xyt$ & $tyz)$,

(96)
$$\forall n \, (x_{i-1}x_{i-1}x_i \ \& \ \neg x_{i-1}x_ix_{i+1}) \ (i = 1, \ldots, 2n \ \&$$
$$\& \ x_{2n+1} = x_0) \Longrightarrow x_{2n}x_0x_1 \,.$$

The characterization of posets by system $\mathbf{P}_3$ has the following meaning. Firstly, the closed betweenness relation of every poset satisfies $\mathbf{P}_3$. Conversely, suppose P is a set endowed with a ternary relation which satisfies $\mathbf{P}_3$. Two elements $x, y \in P$ are said to be *comparable* if xyy or xxy, and *connected* provided they are linked by a *connected sequence*, that is, a sequence $x = x_0, x_1, \ldots, x_n = y$ with the property that any two consecutive terms are comparable. In particular any two comparable elements are connected. The comparability relation is clearly symmetric and also reflexive by (91), hence so is connectedness. Besides, the concatenation of two connected sequences is also connected, again by (91). Therefore connectedness is an equivalence relation, whose cosets will be called the *connected components* P_ι ($\iota \in I$) of P.[†] Now from each connected component P_ι which is not a singleton we choose two distinct comparable elements a_ι, b_ι, Furthermore suppose $x, y \in P$ are distinct comparable elements. Then $x, y \in P_\iota$ for some (unique) $\iota \in I$. Since $b_\iota, x \in P_\iota$, there is a connected sequence $b_\iota, x_1, \ldots, x_n, x$ and this implies that the sequence $a_\iota, b_\iota, x_1, \ldots, x_n, x, y$ is also connected, hence so is the sequence

$$(97) \qquad a_\iota, b_\iota, b_\iota, x_1, x_1, \ldots, x_n, x_n, x, x, y .$$

By replacing all possible occurrences in (97) of three consecutive terms z, z, z by z, z, sequence (97) is reduced to a connected subsequence (possibly the same) of the form

$$(98) \qquad \begin{aligned} a_\iota = z_{r_0}, b_\iota, \ldots, z_{r_1} \; ; \ldots \; ; z_{r_{i-1}}, \ldots, z_{r_i}; \; z_{r_i}, \ldots, z_{r_{i+1}} \; ; \cdots \\ \cdots ; z_{r_s} \ldots, x, z_{r_{s+1}} = y , \end{aligned}$$

where each subsequence $z_{r_{i-1}}, \ldots, z_{r_i}$ consists of distinct terms. One proves that the parity of s does not depend on the choice of the starting sequence $x_1, \ldots, x)n$. This enables one to set

$$(99) \qquad x \leq_{a_\iota, b_\iota |_{\iota \in I}} y \Longleftrightarrow s \equiv 0 \; (\text{mod } 2) .$$

One proves that this relation is a partial order on P. Moreover, the closed betweenness (40) associated with $\leq_{a_\iota b_\iota |_{\iota \in I}}$ coincides with the original ternary relation. On the other hand, if the starting point is a connected poset ($P \leq$), which means there is only one connected component, and the construction (99) is applied to the closed betweenness relation (40), then $\leq_{a_\iota b_\iota |_{\iota \in P}}$ is either the original relation $\leq$ or its dual.

Sholander [1952], using the paper by Altwegg, obtained a shorter system equivalent to $\mathbf{P}_3$, namely

$$\mathbf{P}_4 = \{(100), (101), (102)\} ,$$

where

$$(100) \qquad xyx \Longleftrightarrow x = y ,$$

$$(101) \qquad xyz \; \& \; ytu \Longrightarrow zyt \text{ or } uyx ,$$

$$(102) \qquad \begin{aligned} \forall n \geq 3 \; n \text{ odd } \& \; x_1, \ldots, x_n \text{ distinct } \& \; x_1 x_1 x_2 \; \& \; \ldots \& \; x_{n-1} x_{n-1} x_n \\ \Longrightarrow x_{n-1} x_n x_1 \text{ or } x_n x_1 x_2 \text{ or } \exists i \; 1 \leq i \leq n-2 \; \& \; x_i x_{i+1} x_{i+2} . \end{aligned}$$

[†]This concept of connectedness is borrowed from graph theory.

Morinaga and Nishigōri [1953], unaware of the papers by Altwegg and Sholander, characterized posets by systems

$$\mathbf{P}_5 = \{(91), (95), (103), (104), (105), (107), (108), (109)\},$$

$$\mathbf{P}_6 = \{(91), (103), (104), (105), (107), (108), (110), (111)\},$$

$$\mathbf{P}_7 = \{(91), (103), (105), (106), (107), (112), (113)\},$$

where

(103) $$x \neq z \ \& \ xyz \implies zyx,$$

(104) $$x \neq y \ \& \ xyz \implies xxy,$$

(105) $$x \neq z \ \& \ xyz \ \& \ xzy \implies y = z,$$

(106) $$xyz \text{ or } yzx \text{ or } zxy \implies xyy,$$

(107) $$y \neq z \implies \neg(xyz \ \& \ xzy),$$

(108) $$x \neq z \ \& \ xyz \ \& \ yzt \ \& \ y \neq z \implies xzt,$$

(109)
$$\forall n \ x_1, \ldots, x_{2n+1} \text{ distinct } \& \ x_1 x_2 x_2 \ \& \ x_2 x_3 x_3 \ \& \ldots$$
$$\& \ x_{2n+1} x_1 x_1 \implies \exists i \ 1 \leq i \leq 2n - 1 \ \& \ x_i x_{i+1} x_{i+2},$$

(110)
$$\forall n \ x_1, \ldots, x_{2n+1} \text{ distinct } \& \ x_1 x_2 y_1 \ \& \ x_2 x_3 y_2 \ \& \ldots$$
$$\& \ x_{2n+1} x_1 y_{2n+1} \implies \exists i \ 1 \leq i \leq 2n - 1 \ \& \ x_i x_{i+1} x_{i+2},$$

(111) $$x, y, z, t \text{ distinct } \& \ xyz \ \& \ ytu \implies (xyt \ \& \ \neg tyz) \text{ or } (\neg xyt \ \& \ tyz),$$

(112)
$$x, y, z, t \text{ distinct } \& \ xyz \ \& \ (ytu \text{ or } tuy \text{ or } uyt)$$
$$\implies (xyt \ \& \ \neg tyz) \text{ or } (\neg xyt \ \& \ tyz),$$

(113)
$$\forall n \ x_1, \ldots, x_{2n+1} \text{ distinct } \& \ (x_i y_i x_{i+1} \text{ or } \& \ y_i x_{i+1} x_i \text{ or } x_{i+1} x_i y_i)$$
$$(i = 1, \ldots, 2n + 1 \ \& \ x_{2n+2} = x_1)$$
$$\implies \exists i \ 1 \leq i \leq 2n - 1 \ \& \ x_i x_{i+1} x_{i+2}.$$

The construction and the proofs for $\mathbf{P}_5 - \mathbf{P}_7$ are more complicated than for $\mathbf{P}_4$, but the idea is the same.

The systems $\mathbf{P}_3 - \mathbf{P}_7$ are independent.

Morinaga and Nishigōri [1953] have also characterized partially ordered sets with least element 0 by the independent axiom systems

$$\mathbf{P}_8 = \{(91), (95), (103), (104), (106), (108), (114), (115), (116)\} \, ,$$

$$\mathbf{P}_9 = \{(91), (103), (104), (106), (108), (111), (117)\} \, ,$$

$$\mathbf{P}_{10} = \{(91), (103), (104), (106), (108), (117), (118)\} \, ,$$

where

(114) $$0xx \, ,$$

(115) $$x0y \implies x = 0 \text{ or } y = 0 \, ,$$

(116)
$$x_1, x_2, x_3 \text{ distinct } \& \ x_1 x_2 y_1 \ \& \ x_2 x_3 y_2 \ \& \ x_3 x_1 y_3$$
$$\implies x_1 x_2 x_2 \text{ or } x_2 x_3 x_1 \text{ or } x_3 x_1 x_2 \, ,$$

(117) $$xyz \implies 0xy \text{ or } 0yz \, ,$$

(118) $$x, y, z, t \text{ distinct } \& \ xyz \ \& \ xzt \implies yzt \, .$$

In the same paper the open betweenness relation (14) is used to characterize posets by the system

$$\mathbf{P}_{11} = \{(15), (17), (119), (120)\} \, ,$$

where

(119)
$$x, y, z, t, u \text{ distinct } \& \ (xyz) \ \& \ ((ytu) \text{ or } (tuy) \text{ or } (uyt))$$
$$\implies ((xyt) \ \& \ \neg(tyz)) \text{ or } (\neg(xyt) \ \& \ (tyz)) \, ,$$

(120)
$$\forall n \ x_1, \ldots, x_{2n+1} \text{ distinct } \& \ y_1, \ldots, y_{2n+1} \text{ distinct } \& \ x_{2n+3} = x_2 \ \&$$
$$((x_i x_{i+1} y_i) \text{ or } (x_{i+1} y_i x_i) \text{ or } (y_i x_i x_{i+1})) \ (i = 1, \ldots, 2n + 1 \ \& \ x_{2n+2} = x_1)$$
$$\implies x_i, x_{i+1}, y_i \text{ distinct } (i = 1, \ldots, 2n + 1) \ \&$$
$$\exists j \ 1 \le j \le 2n + 1 \ \& \ (x_j x_{j+1} x_{j+2}) \, ,$$

posets with least element 0 by the system

$$\mathbf{P}_{12} = \{(15), (17), (120), (121)\} \, ,$$

where

(121) $$x, y, z \text{ distinct } \& \ ((xyz) \text{ or } (yzx) \text{ or } (zxy)) \implies (0xy) \text{ or } (0yx) \, ,$$

and posets with 0 and 1 by the system

$$\mathbf{P}_{13} = \{(15), (17), (121), (122)\} \, ,$$

where

(122) x, y, z distinct & $((xyz)$ or (yzx) or $(zxy)) \implies (0x1)$.

The following independent axioms systems for totally ordered sets in terms of closed betweenness:

$$\mathbf{T}_{45} = \{(43), (44), (93), (95)\} \, ,$$

$$\mathbf{T}_{46} = \{(100), (101), (102), (123)\} \, ,$$

$$\mathbf{T}_{47} = \{(41), (100), (101)\} \, ,$$

$$\mathbf{T}_{48} = \{(91), (103), (106), (108), (108), (118), (124), (125)\} \, ,$$

$$\mathbf{T}_{49} = \{(91), (95), (103), (106), (124), (125)\} \, ,$$

were given by Altwegg [1950] ($\mathbf{T}_{46}, \mathbf{T}_{47}$), Sholander [1952] ($\mathbf{T}_{46}, \mathbf{T}_{47}$), and Morinaga and Nishigōri [1953] ($\mathbf{T}_{48}, \mathbf{T}_{49}$,) where we have set

(123) $\forall x \, \forall y \, \forall z \, \exists t \; xty$ or ytz or ztx ,

(124) $x \neq y \implies xyy$,

(125) x, y, z distinct $\implies xyz$ or yzx or zxy .

In a totally ordered set T, the closed betweenness relation has the property that for any $x, y, z \in T$, exactly one of these elements, say $\beta(x, y, z)$, is between the two others. Then the following properties hold:

(126) $\beta(x, x, y) = x$,

(127) $\beta(\beta(x, y, z), \beta(x, y, t), u) = \beta(\beta(z, t, u), x, y)$,

(128) $x \neq \beta(x, y, z) \neq z \implies \beta(x, t, y) = y$ or $\beta(y, t, z) = y$.

Sholander [1952] characterized totally ordered sets in terms of the ternary operation β by the system

$$\mathbf{T}_{50} = \{(126), (127), (128)\} \, ,$$

to the effect that chains satisy $\mathbf{T}_{50}$ and conversely, if a set T endowed with a ternary operation β satisfies $\mathbf{T}_{50}$, then the ternary relation defined by $xyz \iff \beta(x,y,z) = y$ fulfils system $\mathbf{T}_{47}$, hence T becomes a totally ordered set.

The relation of closed betweenness has been also used by Padmanabhan [1966] to characterize totally ordered sets among lattices, as we describe below.

Consider the following predicates in a lattice L:

$$A(a,b,x) \iff a \le x \le b \text{ or } b \le x \le a \ ,$$

$$B(a,b,x) \iff (a \wedge x) \vee (b \wedge x) = x = (a \vee x) \wedge (b \vee x) \ ,$$

$$C(a,b,x) \iff (a \wedge x) \vee (b \wedge x) = x = (a \wedge b) \vee x \ ,$$

$$C'(a,b,x) \iff (a \vee x) \wedge (b \vee x) = x = (a \vee b) \wedge x \ ,$$

$$D(a,b,x) \iff a \wedge b \le x \le a \vee x \ .$$

$A(a,b,x)$ is closed betweenness, denoted by axb in this Appendix. $B(a,b,x)$ was introduced by Glivenko as the lattice-theoretic characterization of metric betweenness and adapted by Pitcher and Smiley as lattice betweenness in arbitrary lattices; cf. Theorems 1.4.2 and 2.2.2. The relations C and C' are characterizations of metric betweenness due to Blumenthal and Ellis.

Remark The following implications are obvious: $A \implies B \implies C \implies D$ and $B \implies C' \implies D$.

Theorem I *The following conditions are equivalent in a lattice L:*
(i) L *is a chain;*
(ii) $D \implies A$.

PROOF: (i)$\implies$ (ii): Suppose the elements a, b, x of a chain L satisfy $a \wedge b \le x \le a \vee b$. Then $a \le x \le b$ or $b \le x \le a$ according as $a \le b$ or $b \le a$.

(ii)$\implies$ (i): Take $a, b \in L$. Since $D(a,b,a)$ is true, it follows that $A(a,b,a)$ also holds, which reduces to $a \le b$ or $b \le a$. □

Theorem II *A lattice L is a chain if and only if conditions A,B,C,C',D are equivalent in L.*

PROOF: Immediate from Theorem I and the Remark.   □

Besides, the same paper provides similar characterizations of modular and distributive lattices within the class of all lattices.

The notation xyz used in this Appendix for closed betweenness coincides with the notation axb in Chapters 1 and 2 for lattice betweenness. In view of Theorem II, this is not so dramatic !

References

ALTWEGG, M.
1950. Zur Axiomatik der teilweise gordneten Mengen. Comment. Math. Helv. 24, 149-155.

BETAZZI, R.
1890.* Teoria delle Grandezze. E. Spoeri, Pisa.

BIRKHOFF, G.
1948. Lattice Theory. Amer. Math. Soc. Coll. Publ., New York.

BURALI-FORTI, C.
1893.* Sulla teoria delle grandezze. Riv. Mat. 3, 76-101.

ČECH, E.
1936. Budove Množiny. Praha.

CLAY, R.E.
1969. Sole axioms for partially ordered sets. Logique Anal. (N.S.) 12, 361-375.

HARARY, F.
1961. A very independent axiom system. Amer. Math. Monthly 68, 159-162.

HAUSDORFF, F.
1914. Grundzüge der Mengenlehre. Leipzig.

HUNTINGTON, E.V.
1904-6. The continuum as a type of order: an exposition of the modern theory. Ann. Math. 6, 151-184; 7, 15-43.
1917. Complete existential theory of the postulates for serial order. Bull. Amer. Math. Soc. 23, 276-289.
1924a. Sets of complete independent postulates for cyclic order. Proc. Nat. Acad, Sci. 10,74-78.

1924b. A new set of postulates for betweenness, with a proof of complete independence. Trans. Amer. Math. Soc. 26, 257-282.
1935. Inter-relations among the principal types of order. Trans. Amer. Math. Soc. 38, 1-9.

HUNTINGTON, E.V.; KLINE, J.R.
1917. Sets of independent postulates for betweenness. Trans. Amer. Math. Soc. 18, 301-325.

HUNTINGTON, E.V.; ROSINGER, K.E.
1932. Postulates for separation of point pairs (reversibility order on a closed line). Proc. Amer. Acad. Arts Sci. (Boston) 67, 61-145.

KATRIÑÁK, T.
1959. Poznámka k usporiadnym množina. Acta Fac. Rer. Natur. Univ. Comenianae (Math.) 4, 291-299.

KEMPE, A.B.
1890. On the relation between the logical theory of classes and the geometrical theory of points. Proc. London Math. Soc. 21, 147-182.

LIHOVÁ, J.
2000. Strict order betweenness. Acta Univ. M. Belii Ser. Math. No.8, 27-33. MR 2002f:06002.

MAC NEILLE, H.M.
1937. Partially ordered sets. Trans. Amer. Math. Soc. 42, 416-460.

MCPHE, J.A.
1962. Axioms for betweenness. J. Math. Soc. 37, 112-116.

MOISIL, GR.C.
1942. Curs de analiză generală. Univ. Bucureşti.

MORINAGA, K.; NISHIGŌRI, N.
1952-3. On the axioms of betweenness. J. Sci. Hiroshima Univ. (A) 16, 177-222; 399-408.

PADMANABHAN, R.
1966. On some ternary relations in lattices. Colloq. Math. 15, 195-198.

PEIRCE, C.S.
1880,87. On the algebra of logic. Amer. J. Math. 3, 15-27; 7, 180-202.

ROSENBAUM, I.
1951. A new system of completely independent postulates for betweenness. Bull. Amer. Math. Soc. 57, 279.

RUSSELL, B,
1903.* Principles of Mathematics. Cambridge.

SAKAI, SH.
1965.* Systems of postulates for a ternary relation characterizing a partial order. Rep. Fac. Sci. Shizuoka Univ. 1, no.1, 1-4. MR 33#7276.

SCHRÖDER, E.
1890-1905. Vorlesungen über die Algebra der Logik. Leipzig; vol.I, 1890; vol.II, 1891, 1905; vol.III, 1895. Reprint: Chelsea Publ. Co. Bronx, New York, 1966.

SHATUNOVSKIĬ, S.O.
1904.* On the postulates defining the concept of magnitude (Russian). Mem. Math. Section Russian Soc. Naturalists (Odessa), 26, 21-25.

SHEPPERD, J.A.H.
1956. Transitivities of betweenness and separation and the definitions of betweenness and separation groups. J. London Math. Soc. 31, 240-248.

SHOLANDER, M.
1952. Trees, lattices, order and betweenness. Proc. Amer. Math. Soc. 3, 369-381.

STOLTZ, O.; GMEINER, J.A.
1901.* Theoretische Arithmetik. Leipzig.

SU, B.
1927.* Geometrical proof of the independence of a system of postulates concerning equaliy and inequality. Tôhoku Math. J. 28, 282-286.

TROMBETTA, M.
1983. A system of axioms for the relation of betweenness (Italian). Rend. Inst. Mat. Univ. Trieste 15, no.1-2, 96-107. MR 86k:06004.

VAILATTI, G.
1892.* Sui principi fundamentali della geometria della retta. Riv. Mat. 2, 71-75.

VAN DE WALLE, W.E.

1924. On the complete independence of the postulates for betweenness. Trans. Amer. Math. Soc. 26, 249-256.

YOONEYAMA, K.

1918. Sets of independent postulates concerning equality and inequality (theory of three undefined relations). Tôhoku Math. J. 14, 171-283.

Appendix C: Quasilattices

Quasilattices were introduced by Padmanabhan [1971]. An algebra $(A, \vee, \wedge)$ of type (2,2) is called a *quasilattice* if both $\vee$ and $\wedge$ are semi-lattice operations such that the natural partial order relation determined by $\vee$ enjoys the substitution property with respect to $\wedge$ and vice-versa; that is, iff

Q1 $x \vee x = x$,

Q2 $x \vee y = y \vee x$,

Q3 $x \vee (y \vee z) = (x \vee y) \vee z$,

Q4 $x \vee y = y \Longrightarrow (x \wedge z) \vee (y \wedge z) = y \wedge z$,

and their duals hold. The class of quailattices is equational, because condition Q4 is equivalent to

Q5 $((x \vee y) \wedge z) \vee (x \wedge z) = (x \vee y) \wedge z$

and dually. For taking $y := x \vee y$ in Q4 we obtain Q5, while if $x \vee y = y$ then Q5 reduces to $(y \wedge z) \vee (x \wedge z) = y \wedge z$. The system of axioms Q consisting of Q1–Q3, Q5 and their duals is independent; most of this fact was proved by Padmanabhan (op. cit.), while the independence of Q3 (and of its dual) was established by Chandran [1979]. The paper by Padmanabhan provides also examples of quasilattices that are not lattices.

Recall (Chapter 1, §3) that an identity $f = g$ is called *regular* if the sets of variables occurring in the two sides of the equation are the same; for instance, any identity valid in a semilattice is regular (cf. Lemma 1.3.2). The identities in system Q are regular and valid in any lattice. Even more, Padmanabhan proved that the set Q implies all the regular identities true in any lattice. In other words, quasilattices capture the essence of regular identities valid in all lattices. The proof is via a representation theorem for quasilattices as the sum of a direct system of lattices. The essential tool is the concept of a partition function in algebras, which is due to Plonka [1968].

The following two theorems are worth mentioning.

The class of quasilattices is the smallest equational class of algebras of type (2,2) including both the class of lattices and the class of semilattices (viewed as algebras $(A, \vee, \wedge)$ with $\vee = \wedge$).

The class of quasilattices is not one-based, but can be defined by two identities.

Exercises

1. Derive

$L_{23} \qquad x \wedge (x \vee y) = x \vee (x \wedge y)$

from Q.

2. A plethora of new axiom systems for lattices are obtained by the following procedure. Let $f = g$ be any non-regular identity true in all lattices. Then $Q \cup \{f = g\}$ is an equational basis for lattice theory.

3. The equivalence between the two distributivity laws $D_1^\wedge$ and $D_1^\vee$ is valid in the class of quasilattices as well.

CHANDRAN, V.R.

1979. A note on Padmanabhan's paper "Regular identities in lattices". Pure Appl. Math. Sci. 10, no.1-2, 13-15.

PADMANABHAN, R.

1971. Regular identities in lattices. Trans. Amer. Math. Soc. 158, 179-188.

PADMANABHAN, R.; PENNER, P.

1999. Structures of free n−quasilattices. Algebra Colloq. 6, 249-260.

PLONKA, J.

1968. Some remarks on sums of direct systems of algebras. Fund. Math. 62, 301-308.

Appendix D: Lukasiewicz-Moisil Algebras

In 1940 Moisil created the algebraic counterparts of the Lukasiewicz 3-valued and 4-valued logics, under the name of Lukasiewicz algebras. Then Moisil introduced the n-valued Lukasiewicz algebras and later on, the θ-valued Lukasiewicz algebras, where θ is the order type of an arbitrary linear order. These algebras have been much studied, both for their own sake and for their interest in logic as well as in switching circuit theory, as shown by Moisil and his school. That is why today we refer to these algebras as Lukasiewicz-Moisil algebras. They include as a particular case the well-known Post algebras, which in their turn generalize Boolean algebras.

A concept also due to Moisil is that of a De Morgan algebra. This term designates a bounded distributive lattice endowed with a unary operation N which satisfies $N(x \vee y) = Nx \wedge Ny$ and $NNx = x$, hence $N(x \wedge y) = Nx \vee Ny$. An n-valued Lukasiewicz-Moisil algebra is a De Morgan algebra equipped with an increasing sequence $\varphi_1 \leq \varphi_2 \leq \cdots \leq \varphi_{n-1}$ of lattice endomorphisms which satisfy

$$\varphi_i(x) \wedge N\varphi_i(x) = 0 \ \ (i = 1, \ldots, n-1) \,,$$

$$\varphi_i(Nx) = N\varphi_{n-i}(x) \ \ (i = 1, \ldots, n-1) \,,$$

$$\varphi_i\varphi_j(x) = \varphi_j(x) \ \ (i, j = 1, \ldots, n-1) \,,$$

$$\varphi_i(x) = \varphi_i(y) \ (i = 1, \ldots, n-1) \Longrightarrow x = y \,.$$

The bibliography compiled below selects papers devoted to the axiomatics of Lukasiewicz-Moisil algebras; most of them refer to 3–valued LM algebras. In particular Petcu [1968] obtained a 4–basis and a 3–basis, while Becchio [1978] defined 3–valued LM algebras in terms of an implication operation. Much more about Lukasiewicz-Moisil algebras can be found in the monograph by Boicescu et al [1991].

ABAD, M.; FIGALLO, A.
1984. Characterization of three-valued Lukasiewicz algebras. Rep.
Math. Logic 18, 47-59.

BECCHIO, D.
1973. Sur les définitions des algèbres trivalentes de Lukasiewicz données
par A. Monteiro. Logique et Analyse 63-64; 339-344.
1978. Logique trivalente de Lukasiewicz. Ann. Sci. Univ. Clermont-
Ferrand 16, 38-89.

BOICESCU, V.; FILIPOIU, A.; GEORGESCU, G.; RUDEANU, S.
1991. Lukasiewicz-Moisil Algebras. North-Holland, Amsterdam.

CIGNOLI, R.
1969. Algebras de Moisil de ordin *n*. PhD Thesis, Univ. Nac. del Sur,
Bahía Blanca.

CIGNOLI, R.; MONTEIRO, A.
1965. Boolean elements in Lukasiewicz algebras. II. Proc. Japan Acad.
41, 676-680.

MOISIL, GR.C.
1940. Recherches sur les logiques non-chrysippiennes. Ann. Sci. Univ.
Jassy 26, 431-466 = [1972], 195-232.
1941a. Notes sur les logiques non-chrysippiennes. Ann. Sci. Univ.
Jassy 27, 86-98 = [1972], 233-243.
1941b. Contributions à l'étude des logiques non-chrysippiennes.
I. Un nouveau système d'axiomes pour les algèbres lukasiewicziennes
tétravalentes. C.R. Acad. Sci. Roumanie 5, 289-293 = [1972], 283-286.
1960. Sur les idéaux des algèbres lukasiewicziennes trivalentes. An.
Univ. C.I. Parhon Bucureşti, Ser. Acta Logica 3, 83-95 = [1972], 244-258.
1972. Essais sur les Logiques Non-Chrysippiennes. Ed. Acad.
R.S.Roumanie, Bucarest.
1972a. Les axiomes des algèbres de Lukasiewicz *n*-valentes. [1972], 288-
310.
1972b. Sur les algèbres de Lukasiewicz *n*-valentes. [1972], 311-324.

MONTEIRO, A.
1963. Sur la définition des algèbres de Lukasiewicz trivalentes. Bull.
Math. Soc. Sci. Math. Phys. R.P.Roumaine (NS) 7(55), 3-12 = Notas de
Lógica Matemática No.21, Inst. Mat., Univ. Nac. del Sur, Bahía Blanca.

MONTEIRO, A.; MONTEIRO, L.

1996. Axiomes indépendants pour les algèbres de Nelson, de Lukasiewicz trivalentes, de De Morgan et de Kleene. In: A. Monteiro, Unpublished Papers. I. Notas de Lógica Matemática No.40, Inst. Mat., Univ. Nac. del Sur (INMABB-CONICET), Bahía Blanca.

MONTEIRO, L.

1963. Axiomes indépendants pour les algèbres de Lukasiewicz trivalentes libres. Bull. Math. Soc. Sci. Math. Phys. R.P.Roumaine (NS) 7(55), 199-202 (1964) = Notas de Lógica Matemática, Inst. Mat., Univ. Nac. del Sur, Bahía Blanca.

1969. Sur le principe de détermination de Moisil dans les algèbres de Lukasiewicz trivalentes. Bull. Math. Soc. Sci. Math. R.P.Roumaine 13(61), 447-448.

PETCU, A.

1968. The definition of the trivalent Lukasiewicz algebras by three equations. Rev. Roumaine Math. Pures Appl. 13, 247-250.

RUDEANU, S.

1994. On the axiomatics of Lukasiewicz-Moisil algebras. An. Şti. Univ. Ovidius (Constanţa) 2, 152-159.

SICOE, C.

1967a. Note asupra algebrelor Lukasiewicz multivalente. Stud. Cerc. Mat. 19, 1203-1207.

1967b. On many-valued Lukasiewicz algebras. Proc. Japan Acad. 43, 725-728.

1967c. A characterization of Lukasiewicz algebras. I. II. Proc. Japan Acad. 43, 729-732; 733-736.

1968. Sur la définition des algèbres lukasiewicziennes polyvalentes. Rev. Roumaine Math. Pures Appl. 13, 1027-1030.

Appendix E: Testing Associativity

It turns out that in the proof of the independence of an axiom system, checking associativity is usually the most difficult point. We list below several papers which provide suggestions in order to facilitate this task.

BOCCIONI, D.

1960. Condizioni di distributività ed associatività unilaterali. Rend. Sem. Mat. Univ. Padova 30, 178-193. MR 22A#4652.

1963a. Condizioni di mutua distributività con ripetizioni. Rend. Sem. Mat. Univ. Padova 33, 60-84. MR 28#50.

1963b. Condizioni independenti ed equivalenti a quella di mutua distributività. Rend. Sem. Mat. Univ. Padova 33, 91-98. MR28#51.

FERRERO, G.; FERRERO COTTI, C.

1975. Come verificare la proprietà associatività. Boll. Un. Mat. Ital. (4), 11, 322-329. MR55#3113.

FRAZER, W.D.

1973. On testing a binary operation for associativity. Combinatorial algorithms (Courant Comput. Sci. Sympos., No.9, 1972), 77-90. Algorithmic Press, New York. MR50#13353.

SZÁSZ, G.

1953. Die Unabhängigkeit der Assoziativitätsbedingungen. Acta Sci. Math. Szeged 15, 20-28. MR15,No.1,p.95.

1954. Über die Unabhängigkeit kommutativer multiplikativer Strukturen. Acta Sci. Math. Szeged 15, 130-142. MR15, p.773.

Appendix F: Complete Existential Theory and Related Concepts

The bibliography below, which has no claim to be exhaustive, collects papers devoted to complete existential theories and complete independence of certain sets of postulates. Most of the papers refer to partially ordered sets, lattices or Boolean algebras. Other fields are ring theory (Gilmer Jr [1966]), binary relations (Petre [2002]), binary operations (Robinson [1971]) and associative semirings (Rudeanu and Vaida [2004]).

The papers by Avann [1964] and Gilmer Jr [1966] are only close to complete existential theories. The notice by Ingraham [1923] comments complete independence in general.

Last but not least, let us recall that the father of complete existential theory is Moore [1910].

AVANN, S.P.
    1964. Dependence of finiteness conditions in distributive lattices. Math. Z. 85, 245-256.

DIAMOND, A.H.
    1933. The complete existential theory of the Whitehead-Russell set of postulates for the algebra of logic. Trans. Amer. Math. Soc. 35, 940-948; correct. ibid. 36(1934), 893.

DINES, L.L.
    1914-5. Complete existential theory of Sheffer's postulates for Boolean algebras. Bull. Amer. Math. Soc. 21(1914-5), 183-188.

DUBREIL-JACOTIN, M.L.; LESIEUR, L.; CROISOT, R.
    1953. Leçons sur la Théorie des treillis, des Structures Algébriques Ordonnés et des Treillis Géométriques. Gauthier-Villars, Paris.

GILMER JR, R.W.
    Eleven nonequivalent conditions on a commutative ring. Nagoya Math. J. 26, 183-194.

HUNTINGTON, E.V.

1917. Complete existential theory for serial order. Bull. Amer. Math. Soc. 23, 276-280.

1924a. Sets of completely independent postulates for cyclic order. Proc. Nat. Acad. Sci. USA 10, 74-78.

1924b. A new set of postulates for betweenness, with a proof of complete independence. Trans. Amer. Math. Soc. 26, 247-282.

INGRAHAM, M.H.

1923. Certain limitations of the value of the complete independence of a set of postulates. Bull. Amer. Math. Soc. 29, 199-200.

MACNEILLE, H.M.

1937. Partially ordered sets. Trans. Amer. Math. Soc. 42, 416-460.

MOORE, E.H.

1910. Introduction to a Form of General Analysis (New Haven Colloq., 1906). New Haven.

PADMANABHAN, R.

1969. Implications among some link axioms in lattice theory. J. Madurai Univ. 1, No.1, 27-40.

PETRE, G.

2002. Teoria complet existenţială a proprietăţilor relaţiilor binare. Gaz. Mat. 20(99), 226-238.

ROBINSON, D.F.

1971. A catalogue of binary relations. New Zealand Math. Mag. 8, 2-11.

ROSENBAUM, I.

1951. A new system of completely independent postulates for betweenness. Bull. Amer. Math. Soc. 57, 279.

RUDEANU, S.

1964. Logical dependence of certain chain conditions in lattice theory. Acta Sci. Math. (Szeged) 25, 209-218.

1973. Elemente de Teoria Mulţimilor. Univ. Bucureşti.

RUDEANU, S.; VAIDA, D.

2004. Semirings in operations research and computer science: more algebra. Fund. Inform. 61, 61-85.

TAYLOR, J.S.

1917. Complete existential theory of Bernstein's set of postulates for Boolean algebras. Ann. Math. 19, 64-69.

1920. Sheffer's set of five independent postulates for Boolean algebras in terms of the operation "rejection" made completely independent. Bull. Amer. Math. Soc. 26, 449-454.

VAN DE WALLE, W.E.

1924. On the complete independence of the postulates for betweenness. Trans. Amer. Math. Soc. 26, 249-356.

Two related concepts were introduced by Church [1925]. Let $\mathbf{A} = \{A_1, \ldots, A_n\}$ be a system of axioms. An axiom of $\mathbf{A}$, say A_1, can be *weakened with respect to* $\mathbf{A}$, if there is an axiom B_1 weaker than A_1 such that the system $\{B_1, A_2, \ldots, A_n\}$ is equivalent to $\mathbf{A}$. Setting $A_2 \& \ldots \& A_n = B$, this means

$$(1) \qquad A_1 \Longrightarrow B_1 \,, \ \ B_1 \not\Longrightarrow A_1 \,, \ \ B_1 \& B \Longrightarrow A_1 \,.$$

It is easy to see that A_1 cannot be weakened with respect to $\mathbf{A}$ iff $\neg A_1 \Longrightarrow B$. For if the latter implication were consistent with (1), then from $B_1 \& B \Longrightarrow A_1$ and $\neg B \Longrightarrow A_1$ we would infer $B_1 \Longrightarrow A_1$. The opposite implication is obtained by taking $B_1 := A_1 \lor \neg B$.

The system $\mathbf{A}$ is said to be *irredundant* if it is independent and no axiom of $\mathbf{A}$ can be weakened with respect to $\mathbf{A}$. Therefore an independent set of postulates is irredundant iff $\neg A_i \Longrightarrow A_j$ for every two distinct axioms of $\mathbf{A}$.

The latter remark shows in particular that irredundancy is inconsistent with complete independence.

For other remarks see Gehman [1926].

CHURCH, A

1925. On irredundant sets of postulates. Trans. Amer. Math. Soc. 27, 318-328.

GEHMAN, H.M.

1926. On irredundant sets of postulates. Bull. Amer. Math. Soc. 32, 159-161.

As is well known, in a model which proves the independence of an axiom A of a system $\mathbf{A}$, the failure of A means that A doesn't hold for certain values of its variables. According to Harary [1961], axiom A is *very independent* if its independence can be proved by a model in which A *never holds*, that is, it fails for all possible values given to its variables. The system $\mathbf{A}$ itself is said to be *very independent* if all of its axioms are so, and *absolutely independent* if for every subset **S** of **A** there is a model in which **A** − **S** holds and each axiom of **S** never holds.

It is proved that the following system of axioms for equivalence relations is absolutely independent: reflexivity, symmetry and transitivity restricted to pairwise distinct elements x, y, z.

HARARY, F.

1961. A very independent axiom system. Amer. Math. Monthly 68, 159-162.

Bibliography

ABBOTT, J.C.
1967a. Implicational algebras. Bull.Math. Soc. Sci. Math. Roumanie 11(59), 3-23.
1967b. Semi-Boolean algebra. Mat. Vesnik 4(19), 177-198.

ARAI, Y.; ISÉKI, K.
1965. Axiom systems of B-algebra. II. Proc. Japan Acad. 41, 908-910.

BENNETT, A.A.
1930. Semi-serial order. Amer. Math. Monthly 39, 289-295.
1933. Solution of Huntington's unsolved problem in Boolean algebra. Bull. Amer. Math. Soc. 39, 289-195.

BERAN, L.
1976. Three identities for ortholattices. Notre Dame J. Formal Logic 17, 251-252.
1982. Boolean and orthomodular lattices - a short characterization via commutativity. Acta Univ. Carolinae Math. Phys. 23, 25-27.

BERNSTEIN, B.A.
1914. A complete set of postulates for the logic of classes, expressed in terms of the opration "exception" and a proof of the independence of a set of postulates due to Del Re. Univ. Calif. Publ. Math. 1, 87-96.
1915-1916. A simplification of the Whitehead-Huntinton set of postulates for Boolean algebras. Bull. Amer. Math. Soc. 22, 458-459.
1916. A set of four independent postulates for Boolean algebras. Trans. Amer. Math. Soc. 17, 50-52.
1924. Complete sets of representatives of two-element algebras. Bull. Amer. Math. Soc. 30, 24-30.
1931. Whitehead and Russell's theory of deduction as a mathematical science. Bull. Amer. Math. Soc. 37, 480-488.

1933. Simplification of the set of four postulates for Boolean algebras in terms of rejection. Bull. Amer. Math. Soc. 39, 783-787.

1934. A set of four postulates for Boolean algebra, in terms of the implicative operation. Trans. Amer. Math. Soc. 36, 876-884.

1936. Postulates for Boolean algebra involving the operation of complete disjunction. Ann. Math. 37, 317-325.

1939. Sets of postulates for Boolean groups. Ann. Math. 40, 420-422.

1944. Postulate-sets for Boolean rings. Trans. Amer. Math. Soc. 55, 393-400.

1950. A dual symmetric definition of Boolean algebra, free from postulated special elements. Scripta Math. 16, 157-161.

BIRKHOF, G.

1948. Lattice Theory. Amer. Math. Soc. Coll. Publ., New York. First edition, 1940. Second edition, 1948; third printing, 1961. Third edition, 1967.

BIRKHOFF, G.; KISS, S.A.

1947. A ternary operation in distributive lattices. Bull. Amer. Math. Soc. 53, 749-752.

BIRKHOFF, G.; VON NEUMANN, J.

1936. The logic of quantum mechanics. Ann. Math. 37, 823=843 = J. von Neumann, Collected Papers. Pergamon Press, 1961, vol.IV, 105-125.

BIRKHOFF, G.D.; BIRKHOFF, G.

1946. Distributive postulates for systems like Boolean algebras. Trans. Amer. Math. Soc. 60, 3-11.

BLUMENTHAL, L.M.; BUMCROT, R.J.

1962. Betweenness relations in lattices. Notices Amer. Math. Soc. 9, 111.

BOOLE, G.

1847. The Mathematical Analysis of Logic. Cambridge.

1854. An Investigation into the Laws of Thought. London. = Reprint, Open Const. Publ. Co, Chicago 1940.

BOSBACH, B.

1969. Komplementäre Halbgruppen. Axiomatik und Arithmetik. Fund. Math. 64, 257-287.

1977. ℓ−group cone and Boolean algebra. A common one-identity axiom. Contributions to Universal Algebra, Colloq. Math. Soc. Janos Bolyai, vol.17. North-Holland, Amsterdam, 41-56.

BURRIS, S.; SANKAPPANAVAR, H.P.
1981. A Course in Universal Algebra. Springer-Verlag, New York.

BYRNE, L.
1946. Two brief formulations of Boolean algebra. Bull. Amer. Math. Soc. 52, 269-272.
1948. Boolean algebra in terms of inclusion. Amer. J. Math. 70, 139-143.
1951. Short formulations of Boolean algebra, using ring operations. Canad. J. Math. 3, 31-33.

CARLOMAN, A.G.
1976. Axiomatica de algebra de Boole. Gac. Mat. (Madrid) (1) 28, No.1-2, 19-24.

CHEN, C.C.; GRÄTZER, G.
1969. On the construction of complemented lattices. J. Algebra 11, 56-63.

CHEREMISIN, A.I.
1958. Solution of a problem of Birkhoff. (Rusian). Uch. Zap. Ivanovsk. Gos. Ped. In-ta 18, 205-207.

CHURCH, A.
1952. Review of C. Pereira's paper "Sobre el álgebra de la lógica de Schröder". J. Symb. Logic 17, 154.

COUTURAT, L.
1911. L'Algèbre de la Logique. II-ème éd. Scientia, Paris.

CROISOT, R.
1951. Axiomatique des lattices distributives. Canad. J. Math. 3, 24-27.

DAHN, B.I.
1998. Robbins algebras are Boolean: A revision of McCune's computer-generated solution of Robbins problem. J. Algebra 208, 526-532.

DAY, A.
1977. Splitting lattices generate all lattices. Algebra Universalis 7, 163-169.

DEDEKIND, R.

1987. Über Zerlegung von Zahlen durch ihre grössten gemeinsamen Teiler. Festchrift Tech. Hochschule Braunschweig, 1-40. = Gesammelte Math. Werke, vol.II, 103-147.

DEL RE, A.

1907. Lezioni sulla Algebra della Logica. Napoli.

1911. Sulla independenza dei postulati dell'algebra della logica. Rend. Acad. Napoli (3) 17, 450-458.

DIAMOND, A.H.

1934a. The complete existential theory of the Whitehead-Huntington set of postulates for the algebra of logic. Trans. Amer. Math. Soc. 35, 940-948; corrections ibid. 36(1935), 893.

1934b. Simplification of the Whitehead-Huntington set of postulates for the algebra of logic. Bull. Amer. Math. Soc. 40, 599-601.

DIAMOND, A.H.; MCKINSEY, J.C.C.

1947. Algebras and their subalgebras. Bull. Amer. Math. Soc. 53, 959-962.

DICKER, R.M.

1963. A set of independent axioms for Boolean algebra. Proc. London Math. Soc. 13, 20-30.

DIEGO, A.; SUAREZ, A.

1966. Two sets of axioms for Boolean algebras. Portugal. Math. 23, 139-145.

DINES, L.L.

1914-15. Complete existential theory of Sheffer's postulates for Boolean algebras. Bull. Amer. Math. Soc. 21, 183-188.

DRAŠIČKOVÁ, H.

1966. [...] Mat.-Fyz.Časopis 16, 1-13.

DUBREIL-JACOTIN, M.-L.; LESIEUR, L.; CROISOT, R.

1953. Leçons sur le Théorie des Treillis, des Structurea Algébriques Ordonnées et des Treilis Géométriques. Gauthier-Villars, Paris.

ELLIS, D.

1949. Notes on the foundations of lattice theory. Publ. Math. Debrecen 1, 205-208.

FELSCHER, W.
1957. Ein unsymmetrisches Assoziativ-Gesetz in der Verbandstheorie. Arch. Math. 8, 171-174.
1958. Beziehungen zwischen Verbands-ähnliche Algebren und geregelten Mengen. Math. Ann. 135, 369-387.

FELSCHER, W.; KLEIN-BARMEN, F.
1959 Zur Axiomatik der kommmutativen Halbverbänden. Arch. Math. 10, 7.

FERENTINOU-NICOLACOPOULOU, J.
1968. Sur une axiomatique des treillis distributifs. Praktika Akad. Athen. 43, 64-69 (1969).

FREESE, R.; JEZEK, J.; NATION, J.B.
1995. Free Lattices. Math. Surveys Monographs #42. Amer. Math. Soc., Providence.

FRINK, O.
1926. The operations of Boolean algebras. Ann. Math. 27, 477-490.
1941. Representations of Boolean algebras. Bull. Amer. Math. Soc. 47, 775-776.

GERRISH, F.
1978. The independence of Huntington's axioms for Boolean algebra. Math. Gaz. 62, no.419, 35-40.

GLIVENKO, V.
1936. Géométrie des systèmes de choses normées. Amer. J. Math. 58, 799-828.

GOETZ, A.
1971. On various Boolean structures in a given Boolean algebra. Publ. Math. Debrecen 18, 103-107.

GOODSTEIN, R.L.
1963. Boolean Algebra. Pergamon Press, Oxford; MacMillan Co., New York.

GOULD, M.I.; GRÄTZER, G.
1967. Boolean extensions and normal subdirect powers of finite universal algebras. Math. Z. 99, 16-25.

GRÄTZER, G.

1962. On Boolean functions. (Notes on lattice theory. II). Rev. Math. Pures Appl. 7, 693-697.

1971. Lattice Theory: First Concepts and Distributive Lattices. Freeman & Co., San Francisco.

1978. General Lattice Theory. Academic Press, New York; 2nd edition, Birkhäuser Verlag, 1998.

1979. Universal Algebra. Second edition, Springer-Verlag, New York-Heidelberg-Berlin. First edition, Van Nostrand Co., Princeton, 1968.

2007. Two problems that shaped a century of lattice theory. Notices Amer. Math. Soc. 54, 696-707.

GRAU, A.A.

1947. Ternary Boolean algebra. Bull. Amer. Math. Soc. 53, 567-572.

GÜTING, R.

1971. Ein neues Axiomensystem für Boolesche Verbände. J. Reine Angew. Math. 251, 212-220.

HAHN, O.

1909. Zur Axiomatik des logischen Gebietkalkul. Arch. Sys. Phil. 15, 345-347.

HAMMER (IVĂNESCU), P.L.

1959. Shefferian algebras. Bull. Math. Soc. Sci. Math. Phys. R.P.Roumaine 3(51), 289-303.

HASHIMOTO, J.

1951. A ternary operation in lattices. Math. Japon. 2, 49-52.

HEDLIKOVÁ, J.; KATRIŇÁK, T.

On a characterization of lattices by the betweenness relation - a problem of M. Kolibiar. Algebra Universalis 28, 389-400.

HENLE, P.

1932. The independence of the postulates for logic. Bull. Amer. Math. Soc. 38, 409-414.

HIGMAN, G.; NEUMANN, B.H.

1952. Groups as groupoids with one law. Publ. Math. Debrecen 2, 215-221.

HOBERMAN, S.: MCKINSEY, J.C.C.
1937. A set of postulates for Boolean algebras. Bull. Amer. Math. Soc. 43, 588-592.

HUNTINGTON, E.V.
1904. Sets of postulates for the algebra of logic. Trans. Amer. Math. Soc. 5, 288-309.
1932. A new set of independent postulates for the algebra of logic, with special reference to Whitehead and Russell's Principia Mathematica. Proc. Nat. Acad. Sci. USA 18, 179-180.
1933a. New sets of independent postulates for the algebra of logic, with special reference to Whitehead and Russell's Principia Mathematica. Trans. Amer. Math. Soc. 35, 274-304. Corrections ibid. 557-558, 971.
1933b. A simplification of Lewis and Langford's postulates for Boolean algebra. Mind 42, 203-207.

IMAI, Y.; ISÉKI, K.
1967. On axioms of B-algebra. Portugal. Math. 26, 23-30.

ISÉKI, K.
1965a. Algebraic formulations of propositional calculi. Proc. Japan Acad. 41, 803-807.
1965b. Axiom systems of B-algebra. Proc. Japan Acad. 41, 808-811.
1965c. A characterization of Boolean algebra. Proc. Japan Acad. 41, 893-897.
1968. A simple characterization of Boolean rings. Proc. Japan Acad. 44, 923-924.
1972. On axioms of Boolean algebra. Proc. Japan Acad. 48, 101-102.

ISÉKI, K.; ÔHASHI, S.
1970. Axiom systems of distributive lattices. Proc. Japan Acad. 46, 409-410.

ISOBE, K.
1973. A note on the definition of a Boolean ring (Japanese). Mem. Kitami Inst. Tech. 4, 321-325.

JIPSEN, P.; ROSE, H.
1992. Varieties of Lattices. Lecture Notes #1553, Springer-Verlag.

KALMAN, J.A.
1959. On the postulates for lattices. Math. Ann. 137, 362-370.

1960. Equational completeness and families of sets closed under subtraction. Indag. Math. 22, 402-405.

1968. A two-axiom definition for lattices. Rev. Roumaine Math. Pures Appl. 13, 669-670.

KALMBACH, G.

1983. Orthomodular Lattices. Academic Press, London/New York.

KATRIŇÁK, T.

1959. Poznámka k usporiadanym množina. Acta Fac. Rer. Nat. Univ. Comenian. 4, 291-299.

1961. O jednej otázke charakterizáce sväzu pomocou ternárnej opéracie. Acta Fac. Rer. Nat. Univ. Comenian. 6, 335-341.

KELLY, D.; PADMANABHAN, R.

1989. Variety-independence in lattice theory. Algebra Universalis 26, 380-394.

2002. Self-dual lattice identities. Algebra Universalis 48, 413-426.

2003. Self-dual bases for varieties of orthomodular lattices. Algebra Universalis 50, 141-147.

2004. Irredundant self-dual bases for self-dual lattice varieties. Algebra Universalis 52, 501-517.

2005. Orthomodular lattices and permutable congruences. Algebra Universalis 53, 227-228.

2007. Another independent self-dual basis for the trivial variety. Algebra Universalis 57, 497-499.

KELLY, L.M.

1952. The geometry of normed lattices. Duke Math. J. 19, 661-669.

KEMPE, A.B.

1890. On the relation between the logical theory of classes and the geometrical theory of points. Proc. London Math. Soc. 21, 147-182.

KIMURA, N.

1950. Independency of the axioms of lattices. Rep. Kōdai Math. Sem. 2, 14.

KLEIN-BARMEN, F.

1932. Über einen Zerlegungssatz in der Theorie abstrakten Verknüpfungen. Math. Ann. 106, 114-130.

1934. Beiträge zur Theorie der Verbände. Math. Z. 39, 227-239.

1939. Axiomatische Untersuchungen zur Theorie der Halbverbände und Verbände. Deutsche Math. 4, 32-43.

1940. Über eine weitere Verallgemeinerung des Verbandsbegriff. Math. Z. 46, 472-480.

1959. Über modifizierte Gruppen- und Verbandsaxiome. Arch. Math. 10, 101-103.

KOBAYASI, M.

1943. On the axioms of the theory of lattices. Proc. Imp. Acad. Tokyo 19, 6-9.

KOLIBIAR, M.

1956a. Charakterizácia sväzu pomocou ternárnej operácie. Mat.-Fys. Časopis Sloven. Akad. Vied 6, 10-13.

1956b. On the axiomatics of modular lattices (Russian). Czechoslovak Math. J. 6(81), 381-386.

1958. Charakterisierung der Verbände durch die Relation "zwischen". Z. Math. Logik u. Grundl. Math. 4, 89-100.

KOLIBIAR, M.; MARCISOVÁ, T.

1974. On a question of J. Hashimoto. Mat. Časopis Sloven. Akad. Vied 24, 179-185.

KUROSH, A.G.

1935. Durchschnittdarstellungen mit irreduziblen Komponente in Ringe und sogenante Dualgruppen. Mat. Sb. 42, 613-616.

LALAN, V.

1950. Equations fonctionnelles dans un anneau booléien. C.R. Acad. Sci. Paris 230, 603-605.

LEWIS, C.I.; LANGFORD, C.H.

1932. Symbolic Logic. New York, London. Second ed., Dover Publ., New York, 1954.

LISOVIK, L.P.

1997. On the axiomatization of Boolean algebras (Russian). Ukr. Mat. Zh. 49, 937-942.

LOWIG, H.F.J.

1969. Review to Ôhashi [1968]. Math. Rev. 39, #1369.

MACNEILLE, H.M.
1937. Partially ordered sets. Trans. Amer. Math. Soc. 42, 416-460.

MALLIAH, C.
1968. Independent postulates of Boolean algebras. Pi Mu Epsilon 4, 378-379.
1971. Short definitions of lattices. Rev. Roumaine Math. Pures Appl. 16, 333-339.
1978. Sets of axioms for distributive lattices. Rev. Roumaine Math. Pures Appl. 23, 187-195.

MARTIN, L.H.
1965. Two ternary algebras and their associated lattices. Amer. Math. Monthly 72, 1088-1091.

MATUSIMA, Y.
1952. On some problems of Birkhoff. Proc. Japan Acad. 28, 19-24.

McCUNE, W.
1997. Solutions of the Robbins problem. J. Automated Reason. 19, 263-276.

McCUNE, W.; PADMANABHAN, R.
1996a. Automated Deduction in Equational Logic and Cubic Curves. Lecture Notes Artificial Intelligence 1095, Springer-Verlag, Berlin.
1996b. Single identities for lattice theory and weakly associative algebras. Algebra Universalis 36, 436-449.

McCUNE, W.; PADMANABHAN, R.; VEROFF, R.
2003. Yet another single law for lattices. Algebra Universalis 50, 165-169.

McCUNE, W.; VEROFF, R.; FITTELSON, B.; HARRIS, K.; FEIST, A.
2002. Short single axiom for Boolean algebra. J. Automat. Reason. 29, 1-16.

McKENZIE, R.N.
1970. Equational bases for lattice theories. Math. Scand. 27, 24-38.

MENDELSOHN, N.S.; PADMANABHAN, R.
1972. A single identity for Boolean groups and Boolean rings. J. Algebra 20, 78-82.
1975. Minimal identities for Boolean groups. J. Algebra 34, 451-457.

MEREDITH, C.A.
1969. Equational postulates for the Sheffer stroke. Notre Dame J. Formal Logic 10, 266-270.

MEREDITH, C.A.; PRIOR, A.N.
1968. Equational logic. Notre Dame J. Formal Logic 9, 212-226.

MILLER, D.G.
1952. Postulates for Boolean algebras. Amer. Math. Monthly 59, 93-96.

MOISIL, GR.C.
1959. Asupra simplificării circuitelor cu tranzistori, a celor cu tuburi electronice şi a celor cu criotroane. Studii Cerc. Mat. 10, 7-67.

MONTAGUE, R.; TARSKI, J.
1954. On Bernstein's self dual set of postulates for the algebra of logic. Proc. Amer. Math. Soc. 5, 310-311.

MOORE, E.H.
1910. Introduction to a Form of General Analysis. New Haven Coll., 1906. Yale Univ. Press, New Haven.

MORGADO, J.
1970. Some characterizations of Boolean rings. An. Acad. Brasil. Ci. 42, 641-644.

NEWMAN, M.H.A.
1941. A characterization of Boolean lattices and rings. J. London Math. Soc. 16, 256-272.

NIEMINEN, U.J.
1976. A ternary operation characterization of semi-Boolean algebras. C.R. Acad. Bulgare Sci. 29, 303-305.

NISHIGŌRI, N.
1954. A note on lattice segments. J. Sci. Hiroshima Univ. (A) 18, 123-127.

ÔHASHI, S.
1968. On definitions of Boolean rings and distributive lattices. Proc. Japan Acad. 44, 1015-1017.

ORE, O.

1935. On the foundations of abstract algebra. I. Ann. Math. 36, 406-437.

PADMANABHAN, R.

1966. On axioms for semi-lattices. Canad. Math. Bull. 9, 357-358.

1966b. On some ternary relations in lattices. Coll. Math. 15, 195-198.

1968. A note on Kalman's paper. Rev. Roumaine Math. Pures Appl. 13, 1149-1152.

1969a. Inverse loops as groupoids with one law. J. London Math. Soc. (2)1, 203-206.

1969b. Two identities for lattices. Proc. Amer. Math. Soc. 20, 409-412.

1971. Regular identities in lattices. Trans. Amer. Math. Soc. 158, 179-188.

1972. On identities defining lattices. Algebra Universalis 1, 359-361.

1977. Equational theory of algebras with a majority polynomial. Algebra Universalis 7, 273-275.

1981. A first order proof of a theorem of Frink. Algebra Universalis 13, 397-400.

1983. A self dual equational basis for Boolean algebras. Canad. Math. Bull. 26, 9-12.

PADMANABHAN, R.; MCCUNE, W.

1995. Single identities for ternary Boolean algebras. Comput. Math. Appl. 29, No.2, 13-16.

PADMANABHAN, R.; MCCUNE, W.; VEROFF, R.

2007. Lattice laws forcing distributivity under unique complementation. Houston J. Math. 33, 391-401.

PADMANABHAN, R.; PENNER, P.

2004. Semilattice operations generated by lattice terms. Order 21, 257-263.

PADMANABHAN, R.; QUACKENBUSH, R.W.

1973. Equational theories of algebras with distributive congruences. Proc. Amer. Math. Soc. 41, 373-377.

PEIRCE, C.S.

1880-84. On the algebra of logic. Amer. J. Math. 3(1880), 15-57; 7(1884), 180-202.

PEREIRA, R.C.
1951. Sobre el álgebra de la lógica de Schröder. Rev. Mat. Hisp. Amer. (Ser. 4) 11, 222-239.

PETCU, A.
1964. Short definitions of lattices, using the associative and absorption laws. Stud. Cerc. Mat. 16, 1265-1280 (Romanian) = Rev. Roumaine Math. Pures Appl. 10(1965), 339-355.
1967. Definirea algebrelor booleene şi a semilaticilor cu ajutorul a două axiome-identităţi. Stud. Cerc. Mat. 19, 891-895.
1971. O caracterizare globală a laticilor şi semilaticilor. Bul. Inst. Petrol, Gaze şi Geol. 27, 265-268.

PHILLIPS, J.D.; VOJTĚCHOVSKÝ, P.
2005. Linear groupoids and the associated wreath products. J. Symbolic Comput. 40, 1106-1125.

PIC, GH.
1969. Contributions à l'axiomatique des treillis distributifs. Studia Univ. Babeş-Bolyai 14, No.1, 3-9.

PICU, C.I.
1982. Caracterizarea laticilor distributive prin echivalenţe sau implicaţii. Stud. Cerc. Mat. 34, 367-369.

PIXLEY, A.F.
1963. Distributivity and permutablility of congruence relations in equational classes of algebras. Proc. Amer. Math. Soc. 14, 105-109.
1971. The ternary discriminator function in universal algebra. Math. Ann. 191, 167-180.

PLOŠČICA, M.
1996. On a characterization of distributive lattices by betweenness relation. Algebra Universalis 35, 249-255.

PONTICOPOULOS, L.
1962. Une contribution à la théorie axiomatique des treillis. Bull. Soc. Math. Grèce 3, 12-18.

POTTS, D.H.
1965. Axioms for semi-lattices. Canad. Math. Bull. 8, 519.

RIEČAN, J.
1957. K axiomatike modulárnych sväzov. Acta Fac. Rer. Nat. Univ. Comenian. 2, 257-262.

RUDEANU, S.
1959. Independent systems of axioms in lattice theory. Bull. Math. Soc. Sci. Math. Phys. R.P.Roumaine 3(51), 475-488.
1961. On the definition of Bolean algebras by means of binary operations (Russian). Rev. Math. Pures Appl. 6, 171-183.
1962. Asupra unor sisteme de postulate ale algebrelor booleene. Com. Acad. R.P.Române 12, 893-899.
1963. Axiomele Laticilor şi ale Algebrelor Booleene. Ed. Acad. R.P.Române, Bucureşti.
1964. Logical dependence of certain chain conditions in lattice theory. Acta Sci. Math. (Szeged) 25, 209-218.
1974. Boolean Functions and Equations. North-Holland, Amsterdam; American Elsevier, New York.

RUDEANU, S.; VAIDA, D.
2004. Semirings in operations research and computer science: more algebra. Fund. Inform. 61, 61-85.

RUEDIN, J.
1966. Sur les groupoïdes distributifs ayant un élément neutre d'un côté et sur l'axiomatique des treillis. C.R. Acad. Sci. Paris 263, 559-562.
1966-67. Groupoïdes distributifs et axiomatique des treillis. Séminaire Dubreil-Pisot. Algèbre et Théorie des Nombres. 20, No.4.
1967. Distributivité et axiomatique des treillis. C.R. Acad. Sci. Paris 265, 812-815.
1967-68. Axiomatique des treillis. Séminaire Dubreil-Pisot. Algèbre et Théorie des Nombres. 21, No.9.
1968. Groupoïdes pseudo-distributifs à gauche et axiomatique des treillis. C.R. Acad. Sci. Paris 266, 399-401.

SALII, V.N.
1988. Lattices with Unique Complements. Transl. Math. Monographs, Amer. Math. Soc., Providence.

SAMPATHKUMAR, E.
1963. A set of minimum number of postulates for a Boolean algebra. Proc. Indian Acad. Sci. 57, 254-258.

SCHRÖDER, E.
1890-1905. Vorlesungen über die Algebra der Logik. Leipzig. Vol.I, 1890; vol.II, 1891, 1905; vol.III, 1895. Reprint Chelsea Publ. Co. Bronx, New York, 1966.

SHEFFER, H.M.
1913. A set of five independent postulates for Boolean algebras, with applications to logical constants. Trans. Amer. Math. Soc. 14, 481-488.

SHOLANDER, M.
1951. Postulates for distributive lattices. Canad. J. Math. 3, 28-30.
1952. Trees, lattices, order and betweenness. Proc. Amer. Math. Soc. 3, 369-381.
1953. Postulates for Boolean algebras. Canad. J. Math. 5, 460-464.

SICOE, C.
1966. Axiom systems of B-algebra. Proc. Japan Acad. 42, 867-870.

SIOSON, F.M.
1964. Equational bases of Boolean algebras. J. Symbolic Logic 29, 115-124.
1965a. On Newman algebras. I. II. Proc. Japan Acad. 41, 31-34; 35-39.
1965b. A remark on Newman algebra. Kyungpook Math. J. 6, 59-67.
1967. Natural equational bases for Newman and Boolean algebras. Compositio Math. 17, 299-310.

SMILEY, M.F.; TRANSUE, W.R.
1943. Applications of transitivites of betweenness in lattice theory. Bull. Amer. Math. Soc. 49, 280-287.

SOBOCIŃSKI, B.
1962. Six new sets of independent axioms for distributive lattices with 0 and 1. Notre Dame J. Formal Logic 3, 187-192.
1972a. Certain sets of postulates for distributive lattices with the constant elements. Notre Dame J. Formal Logic 13, 119-123.
1972b. An abbreviation of Croisot's axiom-system for distributive lattices with 1. Notre Dame J. Formal Logic 13, 139-141.
1972c. A new formalization of Newman algebra. Notre Dame J. Formal Logic 13, 255-264.
1972d. An equational axiomatization of associative Newman algebras. Notre Dame J. Formal Logic 13, 265-269.

1972e. A semi-lattice theoretical characterization of associative Newman algebras. Notre Dame J. Formal Logic 13, 283-285.

1973. A note on Newman's algebra system. Notre Dame J. Formal
Logic 14, 129-133.

1975a. A new postulate-system for modular lattices. Notre Dame J.
Formal Logic 16, 81-85.

1975b. A short postulate-system for ortholattices. Notre Dame J. Formal Logic 16, 141-144.

1976a. A short equational axiomatization of modular ortholattices.
Notre Dame J. Formal Logic 17, 311-316.

1976b. A short equational characterization of orthomodular lattices.
Notre Dame J. Formal Logic 17, 317-320.

1979. Equational two axiom bases for Boolean algebras and some lattice
theories. Notre Dame J. Formal Logic 20, 865-875.

SORKIN, YU.I.
1951. Independent systems of axioms defining lattices (Russian). Ukr.
Mat. Zh. 3, 85-97.

1954. On the embedding of latticoids and lattices (Russian). Dokl.
Akad. Nauk 95, 931-934.

1962. Constructing lattices by three identity-axioms (Russian). Uchen.
Zap. Mosk. Gos. Ped. In-ta No.8 (1962), 61-66.

STABLER, R.E.
1941. Sets of postulates for Boolean rings. Amer. Math. Monthly 48,
28-29.

STAMM, E.
1911. Beitrag zur Algebra der Logik. Montash. Math. 22, 137-149.

STONE, M.H.
1935a. Subsumption of Boolean algebras under the theory of rings.
Proc. Nat. Acad. Sci. 21, 103-105.

1935b. Postulates for Boolean algebras and generalized Boolean algebras. Amer. J. Math. 57, 703-732.

1936. The theory of representations for Boolean algebras and generalized Boolean algebras. Trans. Amer. Math. Soc. 40, 37-111.

SUDKAMP, T.A.
1976. A proof of Sobociński's conjecture concerning a certain set of
lattice-theoretical formulas. Notre Dame J. Formal Logic 17, 615-616.

Szász, G.

1952. On the independence of a postulate system for the distributive lattices. Math. Ann. 124, 291-293.

1963. On independent systems of axioms for lattices. Publ. Math. (Debrecen) 10, 108-115.

Tamura, S.

1970. Axioms for Boolean rings. Proc. Japan Acad. 46, 121-123.

1975. Two identities for lattices, distributive lattices and modular lattices with a constant. Notre Dame J. Formal Logic 16, 137-140.

Tarski, A.

1938. Einige Bemerkungen zur Axiomatik der Boole'schen Algebra. C.R. Séances Soc. Sci. Lettres Varsovie (Class III) 31, 33-35.

1968. Equational logic and equational theories of algebras. Contributions to Mathematical Logic. (Colloq. Hannover 1966), North-Holland, 275-288.

Taylor, J.S.

1917. Complete existential theory of Bernstein's first set of four postulates for Boolean algebras. Ann. Math. 19, 64-69.

1920a. A set of five postulates for Boolean algebras in terms of the operation "exception". Univ. Calif. Publ. Math. 1, 241-248.

1920b. Sheffer's set of five independent postulates for Boolean algebra in terms of the operation "rejection" made completely independent. Bull. Amer. Math. Soc. 26, 449-454.

Trevisan, G.

1951. Su una questione relativa alle strutture distributive. Rend. Accad. Sci. Fis. Mat. Napoli (IV)18, 144-145.

Vaida, D.

1957. Caracterizarea structurilor modulare prin sisteme cu axiome independente. Stud. Cerc. Mat. 8, 457-466.

Van Albada, P.J.

1964. A self-dual system of axioms for Boolean algebra. Proc. Nederl. Akad. Wet. Ser. A 67, 377-381 = Indag. Math. 26, 377-381.

Vassiliou, Ph.

1950. A set of postulates for distributive lattices. Publ. Nat. Tech. Univ. Athens no.5, 3.

VEROFF, R.

2000. Short 2-bases for Boolean algebra in terms of the Sheffer stroke. Tech. Rep. TR-CS-2000-25, Comput. Sci. Dept., Univ. New Mexico. Albuquerque, NM.

2001. Lattice Theory. http://www.cs.unm.edu/~veroff/LT/, 2001.

WANG, SH.CH.

1953. An axiom system for propositional calculus. J. Chinese Math. Soc. 2, 267-274.

WHITEHEAD, A.N.

1898. A Treatise on Universal Algebra, with Applications.Cambridge Univ. Press.

WHITEMAN, A.

1937. Postulates for Boolean algebra in terms of ternary rejection. Bull. Amer. Math. Soc. 43, 293-298.

WIENER, N.

1917. Certain formal invariances in Boolean algebra. Trans. Amer. Math. Soc. 18, 65-72.

WINKER, S.

1992. Absorption and idempotency criteria for a problem in near-Boolean algebras. J. Algebra 153(2), 414-423.

WOOYENAKA, Y.

1951. Remark on a set of postulates for distributive lattices. Proc. Japan Acad. 27, 162-165.

1964. On postulate sets for Newman algebra and Boolean algebra. I.II. Proc. Japan Acad. 40, 76-81; 82-87.

YULE, D.

1926. Zur Grundlegung des Klassenkalkuls. Math. Ann. 95, 446-452.

ZAĬCHIK, A.I.

1974. On the axiomatization of distributive lattices (Russian). Mat. Issled. (Kishinev) 1(31), 143-150.

ZELINKA, B.

1967. Důkaz nezávislosti Birkhoffova systému postulátů pro distributivní svazy s jednotokovým prvkem. Časopis Pěst. Mat. 91, 472-477.

ZYLINSKI, E.

1925. Some remarks concerning the theory of deduction. Fund. Math. 7, 203-209.

Index

(xyz), 161
$-$, 99
$<$, 40
$< a, b, c >$, 37
$B(a, b)$, 33
C, 114
$C(L)$, 71
D, 115
K-segment, 33, 47
$K(a, b)$, 33
N, 114
P, 114
$P_n(S)$, 4
U, 114
$[a, b]$, 30
$\vee$, 1, 7
Φ, 69
Ψ, 69
$\beta(x, y, z)$, 174
$\leq$, 1
$\nabla(\mathbf{K})$, 137
$\nabla_{sd}(\mathbf{K})$, 137
$\rightarrow$, 99
$\tilde{\ }$, 47
$\wedge$, 1, 7
axb, 32
$cn(I)$, 123
$id(\sigma)$, 123
$m(x, y, z)$, 62
n-based, 16
$s(x, y, z)$, 47
x', 71
$\mathrm{Con}(A)$, 122
$\mathbf{L}$, 29
$\mathbf{M}$, 49

$\mathbf{T}$, 29

0, 29
1, 29

Abad, M., 184
Abbott, J.C., 141, 193
absorptio-associative law, 10
absorption identity, 20
Adams, M.E., 117
Altwegg, M., 170, 174, 176
antisymmetry, 1
Arai, Y., 102, 193
arithmetical variety, 125
Avann, S.P., 189

Baker, K.A., 127
basis, 16
Becchio, D., 183
Bennett, A.A., 1, 86, 193
Beran, L., 81, 130, 131, 193
Bergman, G.M., 22
Bernstein, B.A., 74, 75, 77, 86, 94, 95, 100, 101, 103, 104, 193
Betazzi, R., 160, 176
Birkhoff, G., 2, 8, 9, 12, 39, 53, 58, 60, 63, 64, 69, 73, 81, 92, 117, 118, 128, 170, 176, 194
Birkhoff, G.D., 60, 81, 194
Blumenthal, L.M., 33, 175, 194
Boccioni, D., 187
Boicescu, V., 184
Boole, G., 72, 194
Boolean absorption, 117
Boolean algebra, 72

Boolean function, 105
Boolean group, 95
Boolean lattice, 72
Boolean ring, 92
Bosbach, B., 102, 194
bounded lattice, 36
Bumcrot, R.J., 33, 194
Burali-Forti, C., 160, 176
Burris, S., 126, 135, 195
Byrne, L., 88, 94, 114, 195

Carloman, A.G., 84, 195
Cech, E., 163, 176
chain, 160
Chandran, V.R., 181, 182
Chen, C.C., 116, 140, 195
Cheremisin, A.I., 64, 195
Chinese lantern, 129
Church, A., 111, 113, 191, 195
Cignoli, R., 184
Clay, R.E., 176
clone, 3
closed betweenness, 164, 170
closed separation, 169
complement, 71
complementation, 72
complemented lattice, 72
complete existential theory, 74
completely independent, 74
conditional disjunction, 111
congruence, 122
congruence distributive, 125
congruence lattice, 122
conjugate identity, 139
Couturat, L., 114, 195
covering relation, 40
Croisot, R., 2, 11, 60, 65, 107, 189,
    195, 196
cyclic order, 165

Dahn, B.I., 91, 195
Day, A., 139, 140, 195
De Morgan, 71
Dedekind, R., 8, 39, 53, 138, 196
definitional equivalence, 69
Del Re, A., 74, 196

Diamond, A.H., 74, 75, 189, 196
Dicker, R.M., 112, 196
Diego, A., 101, 196
difference, 99
Dilworth variety, 140
Dilworth, R.P., 116
Dines, L.L., 103, 189, 196
disjointness relation, 115
distributive groupoid, 12
distributive lattice, 53
double negation, 71
Drašičková, H., 68, 196
dual axiom, 8
dual theorem, 8
Dubreil-Jacotin, M.-L., 2, 11, 189,
    196

Ellis, D.O., 58, 175, 196
equation, 19
equational class, 9
equational theory, 19
exception, 99
expression, 3

Feist, A., 91, 104, 202
Felscher, W., 3, 11, 12, 58, 197
Ferentinou-Nicolacopoulou, J., 61,
    197
Ferrero Cotti, C., 187
Ferrero, G., 187
field of sets, 93
Fig.2, 53, 54
Figallo, A., 184
Figure 1, 40
Filipoiu, A., 184
finitely definable, 15
Fittelson, B., 91, 104, 202
flock, 144
forbidden-subalgebra theorem, 134
Frazer, W.D., 187
Freese, R., 139, 197
Frink, O., 87, 109, 145, 197
Funayama, N., 123

Gareau, A., 141
Gehman, H.M., 191

generalized Boolean ring, 93
Georgescu, G., 184
Gerrish, F., 74, 197
GGP term, 126
Gilmer Jr, R.W., 189
Glivenko, V., 32, 175, 197
Gmeiner, J.A., 160, 178
Goetz, A., 107, 197
Goodstein, R.L., 88, 197
Gould, M.I., 125, 197
Grätzer, G., 23, 82, 92, 93, 116, 122,
  125, 127, 135, 140, 195, 197
Grau, A.A., 106, 198
greatest lower bound, 1
groupoid, 2
Gueting, R., 101, 198

Hahn, O., 113, 198
Hammer (Ivănescu), P.L., 198
Hammer, P.L., 113
Harary, F., 176, 191
Harris, K., 91, 104, 202
Hashimoto, J., 47, 65, 198
Hasse diagram, 40
Hausdorff, F., 160, 176
Hedliková, J., 35, 198
Henle, P., 86, 198
Higman, G., 20, 198
Hoberman, S., 104, 199
Huntingtom variety, 117
Huntington I, 74
Huntington II, 113
Huntington III, 85
Huntington IV, 85
Huntington property, 117
Huntington V, 85
Huntington VI, 86
Huntington, E.V., 74, 85–87, 100,
  113, 116, 160, 162, 166, 168, 176,
  190, 199

ideal, 123
idempotent, 92
identity, 3
Imai, Y., 102, 199
implication, 99

implication algebra, 141
Ingraham, M.H., 189
interval, 30
irredundant, 191
Iséki, K., 62, 98, 102, 193, 199
Isobe, K., 95, 199

Jónsson term, 21
Jónsson, B., 139
Jezek, J., 139, 197
Jipsen, P., 139, 199
join semilattice, 1

Kalicki, J., 107
Kalman, J.A., 9, 18, 144, 199
Kalmbach, G., 129, 134, 135, 142,
  145, 200
Katriňák, T., 35, 37, 164, 177, 198,
  200
Kelly, D., 36, 39, 48, 50, 51, 79, 134,
  136–138, 200
Kelly, L.M., 47, 200
Kempe, A.B., 177, 200
Kimura, N., 8, 200
Kiss, S.A., 63, 194
Klein-Barmen, F., 2, 3, 9, 11, 197,
  200
Kline, J.R., 162, 177
Kobayasi, M., 9, 201
Kolibiar, M., 33, 35, 42, 46–49, 66,
  201
Kurosh, A.G., 41, 201

Lalan, V., 105, 201
Langford, C.H., 87, 201
lattice, 1
lattice betweenness, 31, 47
lattice-convex set, 33
least upper bound, 1
Lesieur, L., 2, 11, 189, 196
Lewis, C.I., 87, 201
Lihová, J., 177
Lisovik, L.P., 84, 86, 201
Lowig, H.F.J., 62, 138, 201
Lukasiewicz-Moisil algebra, 183

M3, 55
MacNeille, H.M., 114, 177, 190, 202
majority polynomial, 21
Mal'cev term, 135
Mal'cev, A.I., 135
Malliah, C., 10, 67, 89, 202
Marcisová, T., 48, 66, 201
Martin, L.H., 36, 48, 202
Matusima, Y., 12, 60, 202
McCune, W., 16, 19, 23, 91, 104, 108,
    119, 147, 202, 204
McKenzie, R.N., 13, 20, 22, 28, 119,
    127, 139, 202
McKinsey, J.C.C., 104, 196, 199
McNulty, G., 137
McPhee, J.A., 165, 177
median operation, 62
meet semilattice, 1
Mendelsohn, N.S., 96, 202
Meredith, C.A., 91, 104, 144, 203
Miller, D.G., 94, 203
modular lattice, 39
Moisil, Gr.C., 86, 113, 163, 167, 177,
    183, 184, 203
Montague, R., 75, 77, 203
Monteiro, A., 184
Monteiro, L., 185
Moore, E.H., 74, 189, 190, 203
Morgado, J., 98, 203
Morinaga, K., 172–174, 177

N5, 40, 55
Nakayama, T., 123
NAND, 99
Nation, J.B., 139, 140, 197
Neumann, B.H., 20, 198
Newman algebra, 81
Newman, M.H.A., 75, 81, 203
Nieminen, U.J., 112, 203
Nishigōri, N., 30, 41, 172–174, 177,
    203
NOR, 99

O6, 134
O8, 134
Ohashi, S., 62, 199, 203

open betweenness, 161
Ore, O., 1, 8, 39, 41, 204
ortholattice, 81, 129
orthomodular lattice, 129
Otter, 23, 91, 104, 108

Padmanabhan, R., 2, 4, 12, 16,
    18–21, 23, 36, 39, 48, 50, 51, 79, 82,
    87, 96, 108, 119, 127, 134, 136–138,
    141, 175, 177, 181, 182, 190, 200,
    202, 204
paramodulation, 24
partial order, 159
partially ordered set, 160
Peirce, C.S., 1, 53, 113, 160, 177, 204
Peirce operation, 99
Penner, P., 48, 182, 204
Pereira, R.C., 113, 205
permutable congruence(s), 124, 125
Petcu, A., 2, 4, 10, 14, 86, 183, 185,
    205
Petre, G., 189, 190
Phillips, J.D., 91, 205
Pic, Gh., 68, 205
Picu, C.I., 57, 205
Pixley, A.F., 125, 205
Ploščica, M., 35, 205
Plonka, J., 181, 182
polynomial, 3, 15
Ponticopoulos, L., 16, 42, 58, 84, 205
poset, 160
Potts, D.H., 2, 21, 205
principle of duality, 8, 39
Prior, A.N., 91, 144, 203
Problem 1, 170
Problem 64, 64
Problem 65, 60
Problem 7, 12
Prover9, 19, 28, 59, 142, 147

Quackenbush, R.W., 108, 127, 204
quasilattice, 181

reflexivity, 1
regular groupoid, 12
regular identity, 20, 181

rejection, 99
Riečan, J., 42, 206
Robbins' axiom, 91
Robinson, D.F., 189, 190
Rose, H., 139, 199
Rosenbaum, I., 162, 178, 190
Rosinger, K.E., 168, 177
Rudeanu, S., 2, 9, 58, 77, 86, 87, 89,
    92, 105, 107, 184, 185, 189, 190, 206
Ruedin, J., 3, 12, 206
Ruething, D., 74
Russell, B., 178

Sakai, Sh., 178
Salii, V.N., 117–119, 206
Sampathkumar, E., 74, 84, 102, 206
Sankappanavar, H.P., 126, 135, 195
Schröder, E., 8, 53, 72, 113, 160, 178,
    207
segment, 30
self-dual, 8, 37, 39
semidistributive lattice, 139
semilattice, 2
separation, 167
Shatunovskiĭ, S.O., 160, 178
shearing identity, 40
Sheffer function of n variables, 113
Sheffer stroke, 99
Sheffer, H.M., 103, 207
Shepperd, J.A.H., 164, 169, 178
Sholander, M., 59, 62, 65, 68, 75, 95,
    104, 171, 174, 178, 207
Sichler, J., 117
Sicoe, C., 102, 185, 207
signature, 71
simple Boolean function, 105
Sioson, F.M., 77, 81, 207
skew dual, 99
skew lattice, 9
Smiley, M.F., 31, 33, 47, 164, 207
Sobociński, B., 2, 19, 42, 47, 61, 62,
    66, 81, 90, 130, 132, 207
Sorkin, Yu.I., 1, 2, 8, 9, 16, 208
splitting lattice, 139
Stabler, R.E., 93, 98, 208
Stamm, E., 103, 208

Stoltz, O., 160, 178
Stone, M.H., 74, 92, 102, 208
strict partial order, 159
Su, B., 160, 178
Suarez, A., 101, 196
sublattice, 40
subtraction, 144
Sudkamp, T.A., 42, 208
Szász, G., 11, 60, 187, 209

Tamura, S., 18, 19, 46, 61, 62, 94, 209
Tarski, A., 5, 19, 115, 127, 137, 144,
    209
Tarski, J., 75, 77, 203
Taylor, J.S., 100, 101, 104, 190, 209
term, 3, 15
term function, 3
ternary majority decision, 106
ternary rejection, 106
TIT, 137
totally odered set, 160
transitivity, 1
Transue, W.R., 31, 33, 47, 164, 207
Trevisan, G., 65, 209
trivial lattice, 72
Trombetta, M., 178
Tschantz, S., 140
TUT, 137
type of an algebra, 2, 8

Vaida, D., 46, 58, 189, 190, 206, 209
Vailatti, G., 160, 178
Van Albada, P.J., 76, 209
Van de Walle, W.E., 162, 179, 191
variety, 9
Vassiliou, Ph., 64, 209
Veroff, R., 23, 91, 104, 119, 202, 204,
    210
very independent, 191
Vojtěchovský, P., 91, 205
von Neumann, J., 117, 118, 128, 194

Wang, Sh.Ch., 65, 210
Whitehead, A.N., 73, 210
Whiteman, A., 110, 210
Wiener, N., 76, 210

Winker, S., 91, 210   Zaĭchik, A.I., 66, 210
Wolk, B., 4   Zelinka, B., 60, 210
Wooyenaka, Y., 60, 81, 98, 210   Zylinski, E., 105, 210
Wos, L., 24, 91, 104

Yooneyama, K., 160, 179
Yule, D., 116, 210